Modern Mathematics Education for Engineering
Curricula in Europe

Seppo Pohjolainen • Tuomas Myllykoski •
Christian Mercat • Sergey Sosnovsky

Editors

Modern Mathematics Education for Engineering Curricula in Europe

A Comparative Analysis of EU, Russia, Georgia and Armenia

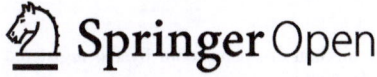 

Editors
Seppo Pohjolainen
Laboratory of Mathematics
Tampere University of Technology
Tampere, Finland

Tuomas Myllykoski
Laboratory of Mathematics
Tampere University of Technology
Tampere, Finland

Christian Mercat
Université Lyon 1
IREM de Lyon, Bâtiment Braconnier
Villeurbanne Cedex, France

Sergey Sosnovsky
Utrecht University
Utrecht, The Netherlands

This project has been funded with support from the European Commission. This publication reflects the views only of the author, and the Commission cannot be held responsible for any use which may be made of the information contained therein.

ISBN 978-3-030-10052-0   ISBN 978-3-319-71416-5   (eBook)
https://doi.org/10.1007/978-3-319-71416-5

Mathematics Subject Classification (2010): 97Uxx

Printed on acid-free paper

This book is published under the imprint Birkhäuser, www.birkhauser-science.com by the registered company Springer International Publishing AG part of Springer Nature.
The registered company address is: Gewerbestrasse 11, 6330 Cham, Switzerland

Preface

Modern science, technology, engineering and mathematics (STEM) education is facing fundamental challenges. Most of these challenges are global; they are not problems only for the developing countries. Addressing these challenges in a timely and efficient manner is of paramount importance for any national economy.

Mathematics, as the language of nature and technology, is an important subject in the engineering studies. Despite the fact that its value is well understood, students' mathematical skills have deteriorated in recent decades in the western world. This reflects in students' slow progressing and high drop-out percentages in the technical sciences.

The remedy to improve the situation is a pedagogical reform, which entails that learning contexts should be based on competencies, engineering students motivation should be added by making engineering mathematics more meaningfully contextualized, and modern IT-technology should be used in a pedagogically appropriate way so as to support learning.

This book provides a comprehensive overview of the core subjects comprising mathematical curricula for engineering studies in five European countries and identifies differences between two strong traditions of teaching mathematics to engineers. It is a collective effort of experts from a dozen universities taking a critical look at various aspects of higher mathematical education.

The two EU Tempus-IV projects—MetaMath (www.metamath.eu) and Math-GeAr (www.mathgear.eu)—take a deep look at the current methodologies of mathematics education for technical and engineering disciplines. The projects aim at improving the existing mathematics curricula in Russian, Georgian and Armenian universities by introducing modern technology-enhanced learning (TEL) methods and tools, as well as by shifting the focus of engineering mathematics education from a purely theoretical tradition to a more application-based paradigm.

MetaMath and MathGeAr have brought together mathematics educators, TEL specialists and experts in education quality assurance from 21 organizations across 6 countries. A comprehensive comparative analysis of the entire spectrum of math courses in the EU, Russia, Georgia and Armenia has been conducted. Its results allowed the consortium to pinpoint issues and introduce several modifications in their curricula while preserving the overall strong state of the university mathematics education in these countries. The methodology, the procedure, and the results of this analysis are presented here.

This project has been funded with support from the European Commission. This publication reflects the views only of the authors, and the Commission cannot be held responsible for any use which may be made of the information contained therein.

This project has been funded with support from the European Commission. This publication reflects the views only of the author, and the Commission cannot be held responsible for any use which may be made of the information contained therein.

Tampere, Finland   Seppo Pohjolainen
Tampere, Finland   Tuomas Myllykoski
Villeurbanne, France   Christian Mercat
Utrecht, the Netherlands   Sergey Sosnovsky

Contents

Contents

About the Editors

Seppo Pohjolainen is a (emeritus) professor at the Laboratory of Mathematics of the Tampere University of Technology. His research interests include mathematical control theory, mathematical modeling and simulation, development and the use of information technology to support learning. He has led several research projects and written a number of journal articles and conference papers on all above-mentioned fields.

Tuomas Myllykoski is a Master of Science and a Teacher of Mathematics at the Laboratory of Mathematics of the Tampere University of Technology. His focus has been on the development and use of learning tools in mathematics, and his current research interests are in the fields of data science, educational psychology and personality.

Christian Mercat is professor in the laboratory Sciences, Société, Historicité, Éducation et Pratiques (S2HEP, EA 4148) and director of the Institute for Research on Mathematics Education (IREM). He trains mathematics teachers at the École Supérieure du Professorat et de l'éducation (ESPE). His main interest is mathematics teaching that respects the creative potential of students. He took part in several research projects on technology-enhanced learning and creative mathematical thinking such as, lately, the mcSquared project (http://www.mc2-project.eu/). He published a number of articles and gave conference talks on the subject, as well as on his main mathematical specialty: discrete differential geometry.

Sergey Sosnovsky is an assistant professor of software technology for learning and teaching at the Institute of Information and Computing Sciences of Utrecht University. He has co-authored more than 90 peer-reviewed publications on topics related to technology-enhanced learning and adaptive information systems. Dr. Sosnovsky served on the programming committees of several conferences and workshops dedicated to Adaptive and Intelligent Educational Technologies. He has built up a strong record of participation in research projects supported by US and EU funding agencies on various aspects of developing AI-based educational technologies, including adaptive support of learning processes and semantic access to instructional resources. Dr. Sosnovsky holds a M.Sc. degree in Information Systems from Kazan State Technological University (Kazan, Russia) and a PhD degree in Information Sciences from University of Pittsburgh (Pittsburgh, PA, USA). He is a receiver of the EU Marie-Curie International Incoming Fellowship. He coordinated the projects MetaMath and MathGeAr, which are at the basis of this book.

Chapter 1
Introduction

1.1 Mathematics Education in EU for STEM Disciplines

Seppo Pohjolainen (✉)
Tampere University of Technology (TUT), Laboratory of Mathematics, Tampere, Finland
e-mail: seppo.pohjolainen@tut.fi

Good competency in mathematics is important in science, technology and economy; mathematics can be considered as the language of nature and technology and also is an important methodology in economics and social sciences. A study by Hanushek and Wößman [21] shows that the quality of education has a strong positive influence on economic growth. In their research, students' skills were measured using 13 international tests, which included mathematics, science, and reading. An OECD report on mathematics in industry [42] states that the remarkable development of the natural sciences and of engineering since the Renaissance is a consequence of the fact that all nature's known laws can be expressed as mathematical equations. The Financial Times outlined their news on February 13th, 2006, as "Mathematics offers business a formula for success".

Despite the fact that the value of mathematics in society and economics is understood, in recent decades students' mathematics skills have deteriorated in western countries. The report "Mathematics for the European Engineer" [50] by the European Society for Engineering Education SEFI[1] states that this phenomenon prevails in Europe. According to the SEFI report, universities in the western world

---

[1] SEFI (http://www.sefi.be).

have observed a decline in mathematical proficiency among new university students and have taken action to remedy the situation. The most common measures are: reducing syllabus content; replacing some of the harder material with more revisions of lower level work; developing additional units of study; establishing mathematics support centres. But sometimes one does nothing.

The decline in mathematical competency may have serious consequences as Henderson and Broadbridge [22] point out. Their message is that industry can only be internationally competitive through mathematical know-how. The number of students majoring in mathematics e.g., in Australia, has decreased, while the number of positions requiring mathematical skills has increased.

The union of Academic Engineers and Architects in Finland[2] (TEK) published a report in 2009 with the recommendation "Knowledge of mathematics and natural sciences must be emphasized more strongly as part of common cultivation and their appreciation should be improved in the society". The report also points out that good command of mathematics and natural sciences is one of the strongest features in engineering studies.

As mathematical proficiency is a prerequisite for studying technical sciences, weak mathematical skills slow down studies. For instance, in Germany the drop-out rate of students sometimes goes up to 35% and one of the primary reasons is the lack of mathematical skills. This caused the industrial Arbeitgeberverband Gesamtmetall to raise an alarm. Drop-out rates in engineering studies are high Europe-wide.[3]

For example, less than 60% of B.Sc. students starting their studies in Finland at Tampere University of Technology (TUT) in 2005 had completed all mandatory first year mathematics courses in four and a half years. Students who had progressed fastest in their studies had typically completed first year mathematics courses according to the recommended schedule. Students who faced problems in studying mathematics more often progressed slowly with their studies in general.

The problems universities are facing with their enrolling students' mathematical proficiency are partly due to school mathematics. The level of school mathematics is being assessed internationally by PISA (The Programme for International Student Assessment), and TIMMS (Trends in International Mathematics and Science Study). PISA is an internationally standardised assessment for 15-year-olds in schools testing literacy in reading, mathematics and science. TIMSS collects educational achievement data at the 4th and 8th grades to provide information on quantity, quality, and content of instruction. The test results from 2012 [44] and 2011 [37] confirm that East-Asian countries are on the top but they are criticised for teacher centred education, large amounts of homework, rote learning etc. EU-countries are doing relatively well, but lagging behind the East-Asian nations, and developing countries can be seen at the bottom.

---

[2]http://www.tek.fi/.

[3]http://ec.europa.eu/information_society/apps/projects/factsheet/index.cfm?project_ref=ECP-2008-EDU-428046.

Learning outcomes in mathematics are not dependent solely on good teaching, sufficient resources or other external considerations with bearing on learning. Factors with bearing on what the student does include attitudes: orientations, intentions and motivations. In order to achieve learning objectives, activity on the part of the learner is required. As student's attitudes and motivational factors are individual, good teaching should take into account student's different learning styles [25].

The recent report 'Mathematics in Europe: Common Challenges and National Policies' by EURYDICE [17] points out that many European countries are confronted with declining numbers of students of mathematics, science and technology, and they face a poor gender balance in these disciplines. The report gives recommendations on how to increase motivation to learn mathematics and encourage the take-up of mathematics-related careers. The report also suggests that the mathematics curriculum should be broadened from contents to competences. Student motivation should be increased by demonstrating and finding evidence how mathematics is used in industry and society, in students' everyday life, and in their future career. New teaching approaches, such as problem-based learning and inquiry-based methods, should be taken into use. Addressing low achievement is important to decrease the drop-out figures, and gender issues should be considered to make mathematics more tempting to female students. Education and professional development of mathematics teachers also plays a key role in this reform.

The European Society for Engineering Education (SEFI), mentioned above, is an international non-profit organisation established in 1973 in Belgium and founded by 21 European Universities. It is an association directly linking the institutions of higher engineering education as an international forum for discussing problems and identifying solutions relating to engineering education. Today, SEFI is the largest network of institutions of higher engineering education, individuals, associations and companies in Europe. Its mission is to contribute to the development and improvement of engineering education in Europe and to the enhancement of the image of both engineering education and engineering professionals in society.

SEFI has set up several working groups on developing engineering education. Among them is the Working Group on Mathematics and Engineering Education, established in 1982. The major outcome of the group resulted in a "Core Curriculum in Mathematics for the European Engineer", first published 1992 and then revised in 2002 as "Mathematics for the European Engineer" [50], and updated 2013 as "A Framework for Mathematics Curricula in Engineering Education" [49].

These documents clearly reflect the European understanding of what the mathematics is that engineers need, and how it should be learned and taught. The 1992 version of the Core Curriculum answers mainly the question: what should be the contents of mathematics courses for engineers? It presented a list of mathematical topics, which are itemised under the headings of Analysis and Calculus, Linear Algebra, Discrete Mathematics and Probability and Statistics.

The SEFI 2002 document identifies four content levels defined as Cores 0, 1, 2, 3. The entry level Core 0 and Core level 1 comprise the knowledge and skills which are necessary and essential for most engineering areas and they should be mandatory for all engineering education, whereas from the other two different parts (Cores 2 and 3) contents will be chosen for the various engineering disciplines. The document also specifies learning outcomes for all the topics and contains additional comments on teaching mathematics.

The most recent report:" A Framework for Mathematics Curricula in Engineering Education", SEFI 2013 [49], proposes a pedagogical reform for engineering mathematics to put more emphasis on what students should know instead of what they have been taught. The learning goals are described as competencies rather than learning contents. Contents should be embedded in a broader view of mathematical competencies that the mathematical education of engineers strives to achieve. Following the Danish KOM project [39], SEFI recommends that the general mathematical competence for engineers is "the ability to understand, judge, do, and use mathematics in a variety of intra- and extra-mathematical contexts and situations in which mathematics plays or could play a role". The general mathematical competence can be divided into eight sub-competencies which are: thinking mathematically, reasoning mathematically, posing and solving mathematical problems, modelling mathematically, representing mathematical entities, handling mathematical symbols and formalism, communicating in, with, and about mathematics, and making use of aids and tools.

Following the SEFI 2013 document we briefly introduce the eight subcompetencies:

Thinking Mathematically This competency comprises the knowledge of the kind of questions that are dealt with in mathematics and the types of answers mathematics can and cannot provide, and the ability to pose such questions. It includes the recognition of mathematical concepts and an understanding of their scope and limitations as well as extending the scope by abstraction and generalisation of results. This also includes an understanding of the certainty mathematical considerations can provide.

Reasoning Mathematically This competency includes the ability to understand mathematical argumentation (chain of logical arguments), in particular to understand the idea of mathematical proof and to understand its the central ideas. It also contains the knowledge and ability to distinguish between different kinds of mathematical statements (definition, if-then-statement, iff-statement etc.). On the other hand it includes the construction of logical arguments and transforming heuristic reasoning into unambiguous proofs (reasoning logically).

Posing and Solving Mathematical Problems This competency comprises on the one hand the ability to identify and specify mathematical problems (pure or applied, open-ended or closed) and the ability to solve mathematical problems with adequate algorithms. What really constitutes a problem is not well defined and it depends on personal capabilities.

Modelling Mathematically This competency has two components: the ability to analyse and work with existing models and to perform mathematical modelling (set up a mathematical model and transform the questions of interest into mathematical questions, answer the questions mathematically, interpret the results in reality and investigate the validity of the model, and monitor and control the whole modelling process).

Representing Mathematical Entities This competency includes the ability to understand and use mathematical representations (symbolic, numeric, graphical and visual, verbal, material objects etc.) and to know their relations, advantages and limitations. It also includes the ability to choose and switch between representations based on this knowledge.

Handling Mathematical Symbols and Formalism This competency includes the ability to understand symbolic and formal mathematical language and its relation to natural language as well as the translation between both. It also includes the rules of formal mathematical systems and the ability to use and manipulate symbolic statements and expressions according to the rules.

Communicating in, with, and about Mathematics This competency includes the ability to understand mathematical statements (oral, written or other) made by others and the ability to express oneself mathematically in different ways.

Making Use of Aids and Tools This competency includes knowledge about the aids and tools that are available as well as their potential and limitations. Additionally, it includes the ability to use them thoughtfully and efficiently.

In order to specify the desired cognitive skills for the topical items and the sub-competences, the three levels described in the OECD PISA document [43] may be used. The levels are: the **reproduction level**, where students are able to perform the activities trained before in the same contexts; the **connections level**, where students combine pieces of their knowledge and/or apply it to slightly different situations; and the **reflection level**, where students use their knowledge to tackle problems different from those dealt with earlier and/or do this in new contexts, so as to have to reflect on what to use and how to use their knowledge in different contexts.

While the necessity of a pedagogical reform is well understood, there still are not enough good pedagogical models scalable to universities' resources available. Modern information and communication technology (ICT) provides a variety of tools that can be used to support students' comprehension and pedagogical reform. Teachers may run their courses using learning platforms like Moodle. In these environments they may distribute course material, support communication, collaboration, and peer learning and organise face-to-face meetings with videoconferencing tools. Students can get feedback on their mathematical skills' from their teacher, peers, and also by using carefully chosen computer generated exercises, which are automatically checked by computer algebra systems (Math-Bridge, System for Teaching and Assessment using a Computer algebra Kernel[4] (STACK)). There

---

[4]http://www.stack.bham.ac.uk/.

exist mathematical programs like MATLAB[5] and Mathematica,[6] which support mathematical modelling of real world problems.

Math-Bridge is one of the learning platforms available to study mathematics. It offers online courses of mathematics including learning material in seven languages: German, English, Finnish, French, Dutch, Spanish and Hungarian. The learning material can be used in two different ways: in self-directed learning of individuals and as a 'bridging course' that can be found at most European universities (Math-Bridge Education Solution[7]).

Internet contains a variety of open source mathematical tool programs (R, Octave, Scilab), computational engines (Wolfram alpha), various visualisations, and apps that can be used alongside the studies. Links to external resources can be easily used to show real world applications. Some universities have started to provide Massive Open Online Courses (MOOC's) available world-wide for their on- and off-campus students.

Information and communication technology can be used to support the learning process in many ways, but great technology cannot replace poor teaching or lack of resources. The use of technology does not itself guarantee better learning results; instead it can even weaken the student performance. This obvious fact has been known for a long time. Reusser [46] among many other researchers, stated that the design of a computer-based instructional system should be based on content-specific research of learning and comprehension and a pedagogical model of the learner and the learning process. In designing computer-based teaching and learning environments real didactic tasks should be considered. One should thoroughly consider what to teach and how to teach. Jonassen [26] has presented qualities of "meaningful learning" that the design and use of any learning environment should meet. The list has been complemented by Ruokamo and Pohjolainen [48].

In a recent report by OECD [41], it was discovered that in those countries where it is more common for students to use the Internet at school for schoolwork, students' average performance in reading declined, based on PISA data. The impact of ICT on student performance in the classroom seems to be mixed, at best. In fact, PISA results show no appreciable improvements in student achievement in reading, mathematics or science in the countries that had invested heavily in ICT for education. One interpretation of these findings is that it takes time and effort to learn how to use technology in education while staying firmly focussed on student learning.

As mathematics is a universal language, the problems in teaching and learning are globally rather similar. The importance of mathematics is internationally well understood and deterioration in the students' skills is recognised. Pedagogical reforms are the way EU is going and pedagogically justified use of information technology and tools will play an important role here.

---

[5]http://mathworks.com.

[6]http://www.wolfram.com/mathematica.

[7]http://www.math-bridge.org.

1.2 TEMPUS Projects MetaMath and MathGeAr

Sergey Sosnovsky
Utrecht University, Utrecht, The Netherlands
e-mail: s.a.sosnovsky@uu.nl

1.2.1 Introduction

The world-wide system of STEM[8] (science, technology, engineering and math) education faces a range of fundamental challenges. Addressing these challenges in a timely and efficient manner is paramount for any national economy to stay competitive in the long range. It is worth noticing that most of these problems are truly global; they are actual not only for the developing countries, but also for countries characterised by stable engineering sectors and successful educational systems (such as, for example, Germany and USA). Experts identify three main clusters of factors characterising the change in the requirements towards the global system of STEM education [40]:

- Responding to the changes in global context.
- Improving perception of engineering subjects.
- Retention of engineering students.

1.2.1.1 Responding to the Changes in Global Context

The speed of renewal of engineering and technical knowledge and competencies is ever growing. Most engineering sectors observe acceleration of the cycle of innovation, i.e. the time between the birth of a technology and its industrial mainstreaming. Technical skills evolve rapidly, new competencies emerge, old competencies dissolve. In 1920, the average half-life of knowledge in engineering was 35 years. In 1960, it was reduced to 10 years. In 1991, it was estimated that an engineering skill is half-obsolete in 5 years. Nowadays, "IT professional would have to spend roughly 10 h a week studying new knowledge to stay current (or upskilling, in the current lingo)" [13]. This essentially means that modern engineering programmes have to teach students how to obtain and operate skills that are not yet defined and how to work at jobs that do not yet exist. Continuous education and retraining is common already and will only widen in the future. This increase in intensity of education and diversity of educational contexts renders the

---

[8]A similar term has been suggested in the German literature as well—MINT ("Mathematik, Informatik, Naturwissenchaft un Technik").

traditional system of STEM education inadequate. New forms of education powered by new educational technologies are required.

At the same time, the engineering and technical problems themselves are changing. Technology has become an integral part of most sectors of a modern economy, technical systems have become more complex and interconnected, even our daily life activities have been more and more penetrated by technology. Finding solutions for this kind of problems and management of this kind of systems requires new approaches that take into account not only their technical side, but also their relations to the social, ecological, economical and other aspects of the problem. Effective teaching of future engineers to deal with these interdisciplinary problems by applying more comprehensive methods should require significant redesign of STEM courses and curriculae. One more group of factors influencing STEM education come from the ever growing globalisation of economic and technological relations between courtiers and companies. Markets and manufacturers have internationalised and the relations between them have become more dynamic. Solutions for typical engineering problems have become a service, basic engineering skills and competencies have transformed into a product that has a market value and can be offered by engineers of other nations. Countries that have not invested in own STEM education and do not possess strong enough engineering workforce will be forced to pay other courtiers for engineering problems by outsourcing.

1.2.1.2   Improving Perception of Engineering Subjects

While the demand for engineer professionals is increasing world-wide, the number of engineering graduates is not growing, and in some countries has even dropped over the last several years. Potential students often consider engineering and technical professions less interesting. Those driven by financial stimuli do not consider engineering as an attractive, money-making career and choose business-oriented majors. Young people motivated rather by a social mission and a public value of their future profession also seldom believe that STEM answers their life goals and remain in such fields as medicine and humanities. Such beliefs are clearly misconceptions as engineering professions are both well paid and societally important. Yet, if the appeal of STEM careers is often not apparent to potential students it is the responsibility of STEM programme administrations and teachers to properly advertise their fields. In addition to that, many students simply consider STEM education too complex and formal, and, at the end, boring. Changing such an image of STEM education is an important practical task that every national system of engineering education must address.

1.2.1.3   Retention of Engineering Students

One of the biggest problems for engineering and science education is the high dropout rates (especially among freshmen and sophomores). For example, in American universities, up to 40% of engineering students change their major to a non-technical one or simply drop out from college [40]. The situation in Europe is similar. For instance, in Germany over the last 15 years, the number of students who do not finish their university programmes has grown by about 10% for most engineering specialisations. Now, depending on the program, this number fluctuates between 25 and 30% nation-wide. For degrees that include an intensive mathematics component, the rate goes up to 40% of all enrolling students [23]. Similar trends can be observed in other developed EU nations (e.g., The Netherlands, Spain, United Kingdom). One of the reasons for this is the fact the traditional structure of engineering education does not enable students to develop their engineer identity for several academic semesters. All engineering programmes start with a large number of introductory "101" courses teaching formal math and science concepts. Only on the second or third year, the actual "engineering" part of the engineering educational programs starts. For a large percentage of students this comes too late. How can engineering students be sooner exposed to the competencies, requirements, and problems and use cases of engineering is a big methodological problem. Should the structure of the engineering curriculum be modified to gradually introduce engineering subjects in parallel with the introductory science courses, or should the structure of the "101" courses be changed to become more attractive to students and include more engineering "flavour", or, maybe, can new educational technologies make these subjects more engaging and help relieve the student retention problem?

1.2.1.4   National Problems of Engineering Education

The three countries addressed in this book (Russia, Georgia and Armenia) have inherited the strong system of school and university STEM education developed in Soviet Union. It was developed to support industrial economy and has used many unique methodological innovations [29]. Yet, after the collapse of Soviet Union, the educational systems of these countries went through a significant transformation. This process has been characterised by several trends, including attempts to resolve the disproportion of the old Soviet education systems that emphasised formal and technical subjects while overlooking the humanities; closing gaps in largely fragmented inherited national educational systems (mostly, the case of Georgia and Armenia); introduction and implementation of elements of the Bologna process and ECTS. Despite the economic and political turmoil of the 1990s, significant progress has been achieved. Yet, there still exist a number of problems impeding further development of these countries' systems of education. This book takes a closer look at the problems pertaining to the mathematics component of STEM education.

1.2.1.5   Role of Mathematics in STEM Education

All the problems mentioned above are especially important when it comes to the mathematics component of STEM education. Math is a key subject for all technical engineering and science programs without exception. In many respects mathematics serves as a lingua franca for other more specialised STEM subjects. The level of math competencies of a student is critical for successful engineering college education, especially in the beginning, when possible learning problems are amplified and math represents a large share of his/her studies. Differences between the requirements of school and university math education can be rather large, especially since different schools can have very different standards of math training. In addition, students themselves often underestimate the volume of math knowledge required to succeed in an engineering university program. In Georgia and Armenia, these problems are especially actual. There is a massive gap between the level of technical competencies of GE/AM school graduates and the requirements they face once enrolling in universities. This gap has emerged as a result of asynchronous reforms of secondary and tertiary education in these countries. According to the statistics of the National Assessment and Examinations Centre of Georgia, about a third of the university entrants fail the national exam. In Armenia, the data of the Ministry of Education and Science shows that the average score of school graduates in math reaches only about 50%. In Russia, the situation is not that drastic. Yet these problems also exist due to unique national circumstances. After the introduction of the unified state exam (USE) in the beginning of the 2000s, Russian universities have abolished the common habit of year-long preparatory math courses that many potential students took. This resulted in a considerable decrease in math competences of freshmen in provincial universities (more prestigious universities of Moscow and St. Petersburg are less affected, as they can select stronger students based on the result of their USE). Finally, from the organisational perspective, a deficit of STEM students coupled with increased market demands for engineering specialists makes many universities loosen enrollment standards, especially with regards to mathematics. Georgia provides a particular example of this situation. Several years ago, when Georgian national tests did not stipulate mandatory subjects, it was a common practice for students with weak school math grades, to not take a math test, yet get accepted to engineering programs. Such a practice not only reduces the overall level of students but also adds an extra load on university teachers. At the end, this will unavoidably result in a decrease in the quality of engineering programme graduates [1].

1.2.2 Projects MathGeAr and MetaMath

The complete titles of the projects are:

- MetaMath: Modern Educational Technologies for Math Curricula in Engineering Education of Russia;
- MathGeAr: Modernisation of Mathematics curricula for Engineering and Natural Sciences studies in South Caucasian Universities by introducing modern educational technologies.

They both have been supported under the 6th call of the Tempus-IV Program financed by the Education, Audiovisual and Culture Executive Agency (EACEA) of EU. The project have been executed in parallel from 01/12/2013 until 28/02/2017.

1.2.2.1 Objectives

MetaMath and MathGeAr projects aimed to address a wide spectrum of the listed problem of math education in engineering programs of Russian, Georgian and Armenian universities. To solve these problems, the projects rely on a comprehensive approach including studying international best practices, analytical review and modernisation of existing pedagogical approaches and math courses. The objectives of the projects include:

- Comparative analysis of the math components of engineering curricula in Russia, Georgia, Armenia and EU and detection of several areas for conducting reforms;
- Modernisation of several math courses within the selected set of programs with a special focus on introduction of technology-enhanced learning (TEL) approaches.
- Localisation of European TEL instrument for partner universities, including digital content localisation with a focus on the introduction of the intelligent tutoring platform for mathematical courses Math-Bridge [51].
- Building up technical capacity and TEL competencies within partner universities to enable the application of localised educational technologies in real courses.
- Pilot evaluation of the modernised courses with real students validating the potential impact of the conducted reform on the quality of engineering education.
- Disseminate results of the projects.

1.2.2.2 Consortia

Projects consortia consisted of organisation from EU and partner countries (Russia for MetaMath and Georgia/Armenia for MathGeAr). The set of EU partners was the same for the two projects and included:

- Universität des Saarlandes—USAAR (Saarbrücken, Germany),
- Université Claude Bernard Lyon I—UCBL (Lyon, France),
- Tampere University of Technology—TUT (Tampere, Finland),
- Deutsches Forschungszentrum für Künstliche Intelligenz—DFKI (Saarbrücken, Germany),
- Technische Universität Chemnitz—TUC (Chemnitz Germany).

Additionally, the MetaMath consortium contained six more partners form Russia:

- the Association for Engineering Education of Russia—AEER (Tomsk, Russia),
- Saint Petersburg Electrotechnical University—LETI (St. Petersburg, Russia),

- Lobachevsky State University of Nizhni Novgorod—NNSU (Nizhni Novgorod, Russia),
- Tver State University—TSU (Tver, Russia),
- Kazan National Research Technical University named after A.N. Tupolev—KNRTU (Kazan, Russia),
- Ogarev Mordovia State University—OMSU (Saransk, Russia).

The MathGeAr consortium, instead of Russian participants, contained five organisation from Georgia and four from Armenia:

- Georgian Technical University—GTU (Tbilisi, Georgia),
- University of Georgia—UG (Tbilisi, Georgia),
- Akaki Tsereteli State University—ATSU (Kutaisi, Georgia),
- Batumi Shota Rustaveli State University—BSU (Batumi, Georgia),
- Georgian Research and Educational Networking Association—GRENA (Tbilisi, Georgia),
- National Centre for Educational Quality Enhancement—NCEQE (Tbilisi, Georgia),
- National Polytechnic University of Armenia—NPUA (Yerevan, Armenia),
- Armenian State Pedagogical University after Khachatur Abovian—ASPU (Yerevan, Armenia),
- Armenian National Centre For Professional Education and Quality Assurance—ANQA (Yerevan, Armenia),
- Institute for Informatics and Automation Problems of the National Academy of Sciences of the Republic of Armenia—IIAP (Yerevan, Armenia).

Partner organisations had different roles in the projects based on their main competencies. Figures 1.1 and 1.2 graphically represent the structure of the two consortia including the main role/expertise for each organisation. This book is a collective effort of all partners from the two consortia.

1.2.2.3   Execution

The projects have been conducted in three main phases. The results of the first phase are essentially the subject of this book. It included the following tasks:

- Development of the methodology for comparative analysis of math courses.
- Pairwise comparative analysis of math courses between EU and partner universities.
- Development of recommendations for the consequent reform of structural, pedagogical/technological/administrative aspects of the target courses.
- Identification of areas where TEL would bring about the most impact, and selection of TEL instruments to lead to this impact.

The second phase built up on the results of the previous one. It included the following activities:

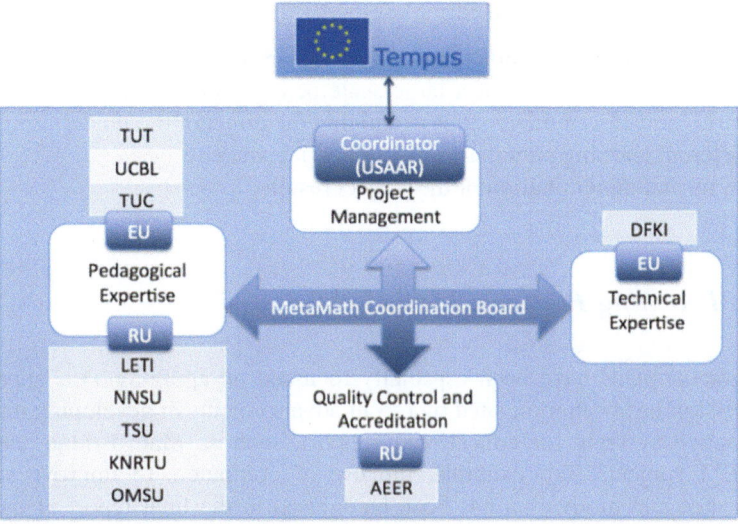

Fig. 1.1 Structure of the MetaMath project consortium

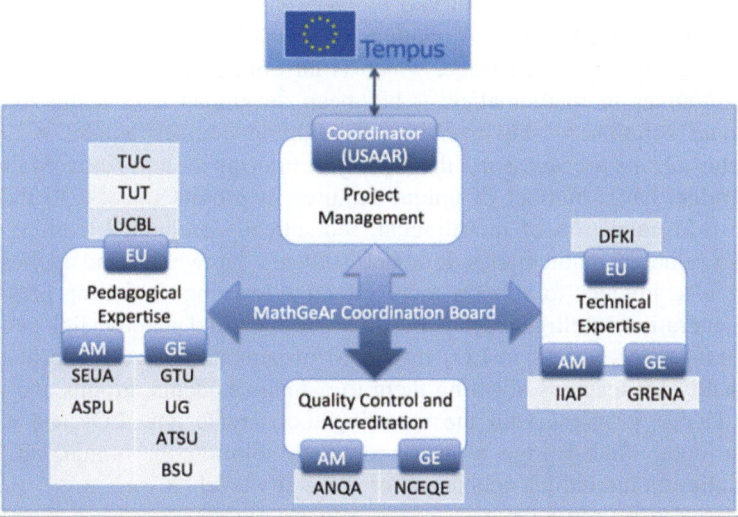

Fig. 1.2 Structure of the MathGear project consortium

- Modification of a set of math courses taught to students of engineering programs in partner universities.
- Localisation of the Math-Bridge platform and its content into Russian, Georgian and Armenian.
- Training of teaching and technical personnel of partner universities to use Math-Bridge and other TEL tools for math education.

Finally, the third phase contained:

- Implementation of the (parts of) modified courses into the Math-Bridge platform.
- Planning and conduction of a large-scale pedagogical experiment across three countries and eleven universities examining the effect of the modernised courses on different learning parameters of engineering students.
- Analysis and dissemination of the project results.

1.2.3   Learning Platform Math-Bridge

Both projects plans have been especially focussed on applying TEL approaches. This decision has been motivated by recent advancements in developing intelligent and adaptive systems for educational support, such as Math-Bridge, especially for STEM subjects. For example, the use of computers to improve students' performance and motivation has been recognised in the final report of the Mathematical Advisory Panel in the USA: "Research on instructional software has generally shown positive effects on students achievement in mathematics as compared with instruction that does not incorporate such technologies. These studies show that technology-based drill and practice and tutorials can improve student performance in specific areas of mathematics" [18]. As the main TEL solution, the projects rely on Math-Bridge, which is an online platform for teaching and learning courses in mathematics. It has been developed as a technology-based educational solution to the problems of bridging courses taught in European universities. As its predecessor—the intelligent tutoring system ActiveMath [36]—Math-Bridge, has a number of unique features. It provides access to the largest in the world collection of multilingual, semantically annotated learning objects (LOs) for remedial mathematics. It models students' knowledge and applies several adaptation techniques to support more effective learning, including personalised course generation, intelligent problem solving support and adaptive link annotation. It facilitates direct access to LOs by means of semantic search. It provides rich functionality for teachers allowing them to manage students, groups and courses, trace students' progress with the reporting tool, create new LOs and assemble new curricula. Math-Bridge offers a complete solution for organizing TEL of mathematics on individual, course and/or university level.

1.2.3.1   Math-Bridge Content

The Math-Bridge content base consists of several collections of learning material covering the topics of secondary and high school mathematics as well as several university-level subjects. They were originally developed for teaching bridging courses by mathematics educators from several European universities. Compared to the majority of adaptive e-learning applications, Math-Bridge supports a mul-

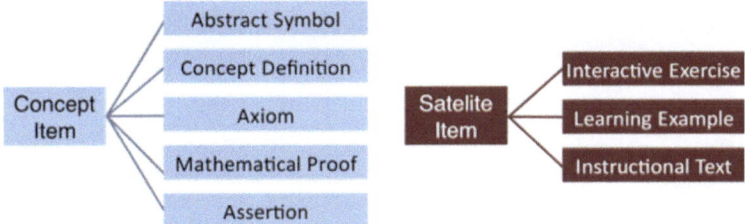

Fig. 1.3 Hierarchy of LO types in Math-Bridge

titude of LO types. The OMDoc language [30] used for representing content in Math-Bridge defines a hierarchy of LOs to describe the variety of mathematical knowledge. On the top level, LOs are divided into concept objects and satellite objects. Satellite objects are the main learning activities; they structure the learning content, which students practice with: exercises, examples, and instructional texts. Concept objects have a dualistic nature: they can be physically presented to a student, and she/he can browse them and read them; at the same time, they are used as elements of domain semantics, and, as such, employed for representing knowledge behind satellite objects and modelling students' progress. Figure 1.3 provides further details of the types of LOs supported in Math-Bridge.

1.2.3.2   Learning Support in Math-Bridge

The Math-Bridge platform provides students with multilingual, semantic and adaptive access to mathematical content. Its interface consists of three panels (Fig. 1.4). The left panel is used for navigation through learning material using the topic-based structure of the course. The central panel presents the math content associated with the currently selected (sub)topic. The right panel provides access to the details of the particular LO that a student is working with, as well as some additional features, such as semantic search and social feedback toolbox.

Math-Bridge logs every student interaction with learning content (e.g., loading a page or answering an exercise). The results of interactions with exercises (correct/incorrect/partially correct) are used by the student-modelling component of Math-Bridge to produce a meaningful estimation of the student's progress. For every math concept the model computes the probabilities that the student has mastered it. Every exercise in Math-Bridge is linked with one or several concepts (symbols, theorems, definitions etc.) and the competencies that the exercise is training for these concepts. A correct answer to the exercise is interpreted by the system as evidence that the student advances towards mastery and will result in the increase of corresponding probabilities. Math-Bridge implement three technologies for intelligent learning support:

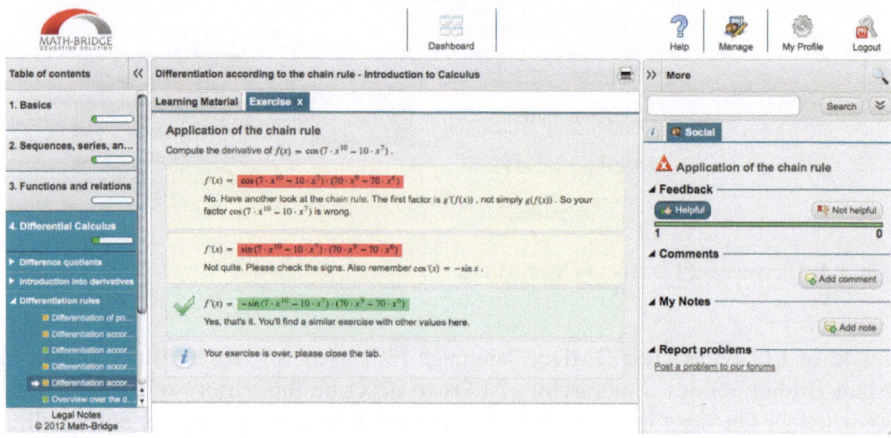

Fig. 1.4 Math-Bridge student interface

- **Personalised courses.** The course generator component of Math-Bridge can automatically assemble a course optimised for individual students' needs and adapted to their knowledge and competencies.
- **Adaptive Navigation Support.** Math-Bridge courses can consist of thousands of LOs. The system helps students find the right page to read and/or the right exercise to attempt by implementing a popular adaptive navigation technique— adaptive annotation [9]. The annotation icons show the student how much progress she/he has achieved for the corresponding part of learning material.
- **Interactive Exercises and Problem Solving Support.** The exercise subsystem of Math-Bridge can serve multi-step exercises with various types of interactive elements and rich diagnostic capabilities. At each step, Math-Bridge exercises can provide students with instructional feedback, ranging from mere flagging the (in)correctness of given answers to presenting adaptive hints and explanations.

1.2.4   Conclusion

Now that both MetaMath and MathGeAr have finished, it is important to underline that their success was in many respects dependent on the results of the comparative analyses conducted during their first phase. Although the overall project approach was defined in advance, individual activities have been shaped by the findings of projects partners contrasting various aspects of math education in Russia/Georgia/Armenia and the EU. The rest of this book presents these findings in detail.

1.3 Perceptions of Mathematics

Christian Mercat and Mohamed El-Demerdash and Jana Trgalova
IREM Lyon, Université Claude Bernard Lyon 1 (UCBL), Villeurbanne, France
e-mail: christian.mercat@math.univ-lyon1.fr; jana.trgalova@univ-lyon1.fr

Pedro Lealdino Filho
Université Claude Bernard Lyon 1 (UCBL), Villeurbanne, France
e-mail: pedro.lealdino@etu.univ-lyon1.fr

1.3.1 Introduction

The global methodology of this comparative study project is based on the analysis of the proposed curriculum and of the actual way this curriculum is implemented in the classroom, in order to identify venues for improvements and modernisations, implement them and study their effect.

In the literature, the philosophical features of the scientific spirit are evident in the sciences which need more objectiveness, chiefly mathematics [11]. From the ingenuous perception of a phenomenon, a pre-scientific spirit needs to overcome a set of epistemological obstacles to reach a scientific stage. We consider this scientific stage an important factor in order to learn, acquire and improve mathematical competencies. The definition of mathematical competence on this project follows the one used in the Danish KOM project and adapted in the SEFI Framework. It is defined as "the ability to understand, judge, do, and use mathematics in a variety of intra- and extra-mathematical contexts and situations in which mathematics plays or could play a role". The attitude towards mathematics is a long standing strand of research and uses reliable measuring tools such as the seminal ATMI [52]. But our research identified dimensions which we find specific for engineering, especially through its relationship with reality.

Mathematics is considered as the foundation discipline for the entire spectrum of Science, Technology, Engineering, and Mathematics (STEM) curricula. Its weight in the curriculum is therefore high [1]. In Armenia, Georgia and Russia, all university students pursuing this kind of curriculum are obliged to take a three semester standard course in higher mathematics. Special studies in Europe suggest that a competencies gap in mathematics is the most typical reason for STEM students to drop out of study [4, 5, 10, 24, 31].

Several research studies show that students' perceptions of mathematics and of mathematics teaching have an impact on their academic performance of mathematics [12, 38], and a positive attitude and perceptions toward the subject will encourage an individual to learn the subject matter better. In a broader sense, perceptions towards mathematics courses are also important to take into account in order to grasp the *cultural differences* between all the different institutions, from the point of view of the students of the course, of their teachers and of the engineers themselves.

The fact that culture does influence these beliefs, while seemingly obvious, is not widely studied. Therefore this study fills a gap in the literature. Without assuming cultural determination, we do show significant differences between institutions. We present in this section a study that investigates these issues, the methodology of data acquisition, the main themes that the study investigates, and the main results.

In order to evaluate students' perceptions of mathematics we elaborated an online survey spread out over all participant countries (Armenia, France, Finland, Georgia and Russia). The survey was elaborated to investigate the three main dimensions of the mathematical courses:

- The usefulness of mathematics.
- Mathematical courses in engineering courses—Contents and Methods.
- Perception of Mathematics.

The survey was spread with a web tool and translated into each partner language to ensure that the meaning of the questions was adequately taken into account. A total of 35 questions were answered by 1548 students from all participant countries.

After collecting the data from the online survey we used the statistical package R to analyse the data and draw conclusions.

We performed a Principal Component Analysis to verify whether there were some patterns in the students' responses. Using the graphical representation of data, we can propose the hypothesis that the methodology of teaching mathematics of each country shapes the average students' perception towards mathematics. Thus, the first conclusion of this analysis is that in the European universities the mathematics are taught as tools to solve problems, that is, mathematics by practicing, while in the non-European universities, the mathematics are taught focussing on proofs and theorems, that is, mathematics by thinking.

1.3.2 *Theoretical Background and Research Question*

Furinghetti and Pehkonen [19] claim that students' beliefs and attitudes as regards mathematics have a strong impact on their learning outcomes. Mathematics-related perceptions are referred to as a belief system in the literature [15, 19]. Furinghetti and Pehkonen point out that there is a diversity of views and approaches to the study of beliefs in the field of mathematics education, and they conclude that the definition of the concept of belief itself remains vague. Some researchers acknowledge that beliefs contain some affective elements [35], while others situate beliefs rather on the cognitive side [53]. Furinghetti and Pehkonen (ibid.) bring to the fore a variety of concepts used by researchers to address issues related to beliefs, such as conceptions, feelings, representations or knowledge. Some authors connect these concepts, for example beliefs and conceptions, as Lloyd and Wilson [34]: "We use the word conceptions to refer to a person's general mental structures that encompass knowledge, beliefs, understandings, preferences, and views". Others distinguish clearly between the two concepts, as Ponte [45]: "They [beliefs] state that something is either true or false, thus having a propositional nature. Conceptions are cognitive constructs that may be viewed as the underlying organizing frame of concepts. They are essentially metaphorical." Furinghetti and Pehkonen (ibid.) attempt to relate beliefs or conceptions and knowledge by introducing two aspects of knowledge: "objective (official) knowledge that is accepted by a community and subjective (personal) knowledge that is not necessarily subject to an outsider's evaluation" (p. 43).

The purpose of this study is not to contribute to the theoretical discussion on these concepts, but rather to study how engineering students view mathematics and its teaching in their schools. We therefore adopt the word 'perceptions' to address these students' views and opinions.

Breiteig, Grevholm and Kislenko [8] claim that "there are four sets of beliefs about mathematics:

- beliefs about the nature of mathematics,
- beliefs about teaching and learning of mathematics,
- beliefs about the self in context of mathematics teaching and learning,
- beliefs about the nature of knowledge and the process of knowing."

Our interest has thus been oriented toward such perceptions of mathematics found with students in engineering courses. These students, engaged in the sciences, have nevertheless different positions, whether philosophical, practical or epistemological, towards mathematics.

The study thus investigates the following question:

"How far do the students' perceptions of mathematics in engineering courses regarding the usefulness of mathematics in real life, the teaching of mathematics (contents and methods) and the nature of mathematics knowledge differ in terms of university, country (France, Finland, Russia, Georgia, Armenia), region (Caucasian, European, Russian) and gender (female, male)?"

1.3.3   Method and Procedures

Drawing on prior studies [16, 20, 38] related to students' mathematics perceptions, and in particular the four sets of beliefs about mathematics suggested by Breiteig et al. [8], we have designed a questionnaire to gather and assess students' perceptions of mathematics and their mathematics courses, and to get concrete indicators of their beliefs about the following:

1. Usefulness of mathematics.
2. Teaching of mathematics in engineering schools, its contents and methods.
3. Nature of mathematical knowledge.

Given the target audience, namely students in engineering courses, we decided not to address beliefs about 'the self in context of mathematics teaching and

Table 1.1 Questionnaire dimensions and numbers of corresponding items

Questionnaire dimension	Number of questions
1. Usefulness of mathematics	8
2. Teaching of mathematics in engineering schools, contents and methods	15
3. Nature of mathematics knowledge	12
Total	35

Table 1.2 Items related to the dimension "usefulness of mathematics"

Item n	Usefulness of mathematics
1	Mathematics has an interest, especially for solving concrete problems.
2	Mathematics is used mostly in technical domains.
3	Mathematics is useless in everyday life.
4	Only applied mathematics is interesting.
5	Mathematics serves no purpose in human sciences.
6	Natural phenomena are too complex to be apprehended by mathematics.
7	Mathematics can be applied to man crafted objects and much less to objects found in nature.
8	Learning mathematics in early classes serves mostly the purpose to help children get around in life.

learning' because this perception is not our focus here: mathematics cannot be avoided and has to be confronted with.

Based on the three above-mentioned dimensions of the questionnaire, we developed 35 questions that cover these dimensions as shown in Table 1.1.

The first dimension of the questionnaire (Table 1.2) explores the students' beliefs about the usefulness of mathematics. Chaudhary [14] points out that mathematics is useful in everyday life: 'since the very first day at the starting of the universe and existence of human beings, mathematics is a part of their lives' (p. 75) and in some professions, such as architects who 'should know how to compute loads for finding suitable materials in design', advocates who 'argue cases using logical lines of reasons; such skill is developed by high level mathematics courses', biologists who 'use statistics to count animals', or computer programmers who develop software 'by creating complicated sets of instructions with the use of mathematical logic skills' (p. 76). In line with such a view of the utility of mathematics, we proposed eight items addressing the utility of mathematics in everyday life (items 1, 3, 8), as well as in technical (item 2), natural (items 6, 7) and human (item 5) sciences. Item 4 questions the perception of the usefulness of mathematics in relation with the distinction between pure and applied mathematics.

A higher education evaluation conducted in France in 2002 [6] focussed, among others, on the mathematics teaching in engineering schools. The study concludes that mathematics takes a reasonable place among the subject matters taught: the amount of mathematics courses is 16% of all courses in the first year of study, in the second year 10% and in the third one only 6%. Another result of this study

Table 1.3 Items related to the dimension "teaching mathematics in engineering schools"

Item n	Teaching mathematics in engineering schools
9	In engineering school, mathematics are mostly pure and abstract.
10	We need more applied mathematics in engineer training.
11	In engineering school, theory are taught without taking into account their applications.
12	There is almost no connection between math teaching and the engineer job reality.
13	Math teaching does not try to establish links with other sciences.
14	Mathematics weigh too much in engineer training.
15	Mathematics cannot be avoided in engineer training.
16	A teacher's only purpose is to bring knowledge to students.
17	The structure of math courses does not allow learning autonomy.
18	With new means available to students, learning is no longer required; one just has to quickly find solutions to problems that are encountered.
19	Courses have not changed in the last decades when the world is evolving greatly and fast.
20	The mathematics courses are extremely theoretic.
21	The mathematics courses are not theoretical enough.
22	The mathematics courses are extremely practical.
23	The mathematics courses are not practical enough.

pointed out that engineering students are mostly taught basic mathematics and do not encounter enough applications. Based on these results, we wished to gather students' perceptions about the teaching of mathematics they are given in their engineering schools. Thus, in the second dimension of the questionnaire (Table 1.3), we decided to address the students' perceptions of the place mathematics takes in their engineering education (item 14 and 15), of the balance between theoretical and practical aspects of mathematics (items 9, 10, 11, 20, 21, 22, and 23) and of the links established between mathematics and other subject matters (item 13). In addition, we wanted to know whether the students feel that the mathematics teaching they receive prepares them for the workplace (item 12). Finally, a set of items addressed the students' perceptions of the methods of teaching of mathematics (items 16, 17, 18, and 19).

The third dimension attempted to unveil the students' implicit epistemology of mathematics, i.e. their perceptions of "what is the activity of mathematicians, in what sense it is a theoretical activity, what are its objects, what are its methods, and how this all integrates with a global vision of science including the natural sciences" [7]. The items related to this dimension (Table 1.4) addressed the relations of mathematics with reality (items 27 and 30), with the truth (items 28, 29 and 32), with other sciences (item 25), with creativity (items 24 and 26), and with models (item 31). Moreover, they questioned the students' perceptions of the nature of mathematical objects (item 35) and the accessibility of mathematical knowledge (items 33 and 34).

Table 1.4 Items related to the dimension "perception of mathematics"

Item n	Perception of mathematics
24	In mathematics, there is nothing left to discover.
25	Mathematics are only a tool for science.
26	There is no room for creativity and imagination in mathematics.
27	Mathematics raised from concrete needs.
28	There is no room for uncertainty in mathematics.
29	Only math can approach truth.
30	Mathematics is only an abstraction, it does not deal with reality.
31	A mathematical model is necessarily limited.
32	A mathematical theory cannot be refuted.
33	Mathematics cannot be the subject of a conversation (contrarily to literature or philosophy).
34	Mathematics is better left to experts and initiated people.
35	Mathematics is a human construction.

Table 1.5 Participants of the study

Country	Number of students	Number of students with completed responses
Armenia	24	12
Finland	189	112
France	430	245
Georgia	285	179
Russia	612	410
Total	1548	958

We calculated the reliability coefficient (Cronbach's alpha) by administering an online version of the questionnaire to a sample of 1548 students from all participant countries (see the Sample section). Students' responses were analysed to calculate the scores of each student. The reliability coefficient (Cronbach's alpha) of 0.79 is high enough to consider our questionnaire as a whole construct a reliable measuring tool.

The experimental validity of the questionnaire as an estimation of the tool validity is also calculated by taking the square root of the test reliability coefficient [3]. Its value of 0.89 shows that the questionnaire has a high experimental validity.

An operational definition of students' perception of their engineering courses is therefore defined for us as a random variable taking vector values represented by the Likert score of the students on the 35 items of the prepared questionnaire on a 1–6 Likert-type scale.

The population on which we base this study are students from partner universities in two Tempus projects, MetaMath in Russia and MathGeAr in Georgia and Armenia, and French and Finnish students on the European side. Within this population a sample of 1548 students filled in the survey with 958 complete responses (incomplete surveys discarded)—see Table 1.5.

1.3.4   Data Analysis

After collecting the data from the online survey we used the statistical package R to analyse the data and draw preliminary conclusions. We performed a Principal Component Analysis (PCA) [27, 47, 54] to investigate patterns in the students' responses. Although the students' responses are not strictly speaking continuous but are a Likert scale between 1 and 6, Multiple Factor Analysis, where different Likert values are not numerically linked but used as simply ordered categories, did not yield finer results. PCA uses a vector space transform to reduce the dimensionality of large data sets giving some interpretation to variability. The original data set, which involves many variables, can often be interpreted by projecting it on a few variables (the principal components).

We used PCA to reveal patterns in students' responses. Using the two first principal components, explaining almost a quarter of the variability, we identify the main common trends and the main differences. Principal components are described in Table 1.6 and Fig. 1.5. In particular, the main result is that we can verify the hypothesis that the methodology of teaching mathematics of each partner, and in particular each country, shapes the average students' perception of mathematics.

During students' interviews and study visits in the project, we could point out the main trends in the way mathematics is taught in the partner institutions. Thus it appeared that mathematics in Europe is taught as a sophisticated tool addressing

Table 1.6 Importance of the components

	PC1	PC2	PC3	PC4
Standard deviation	2.3179	1.69805	1.39846	1.2321
Proportion of variance	0.1535	0.0823	0.05588	0.0433
Cumulative proportion	0.1535	0.2358	0.29177	0.3351

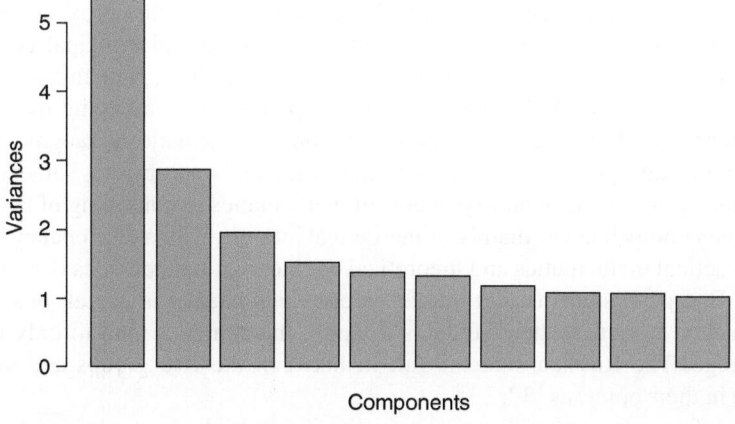

Fig. 1.5 Variance of the first 10 principal components

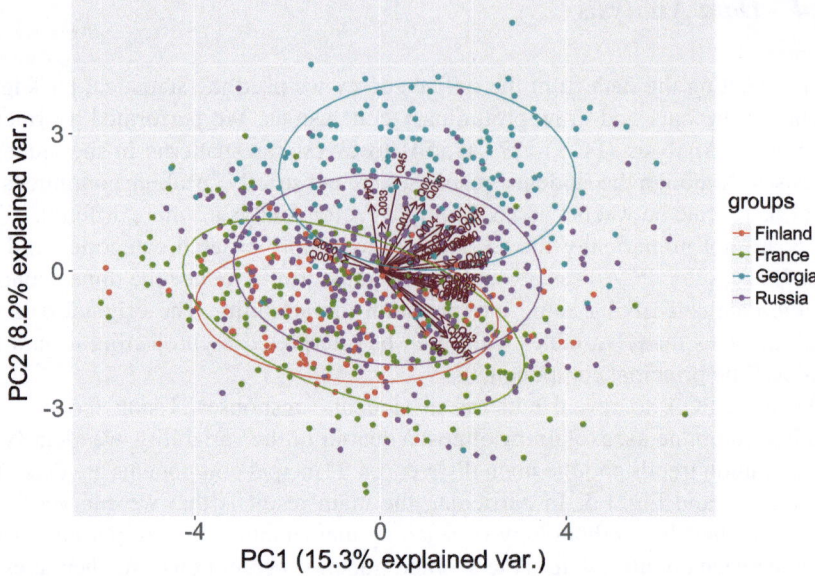

Fig. 1.6 PCA grouped by countries

real engineer's issues; it stands out with respect to a more theoretical approach in the East. This fact does show in the data.

The analysis shows that all engineering students responding to the questionnaire (15.2% of the variability) feel that mathematics teaching is too theoretical, is not practical enough and does not have enough connection with other sciences and the reality of an engineer's job. Therefore, modernised curricula for engineers should address these issues. On the other hand, we identify that Finnish and French students (Fig. 1.6) share most of their perceptions, while the Caucasian students notably differ from them, the Russian students lying in between with a broader variability even given their size. The semantic analysis of the second principal component (8.6% of variability) reveals that in the European universities, mathematics is taught as a tool to solve problems, that is to say, by practicing of applying mathematics to problems, while in the Caucasian universities, mathematics is taught focussing on theorems and proofs, that is to say, mathematics is an abstract subject matter. The Caucasian students tend to perceive of mathematics as consisting of knowledge rather than competencies, mainly of theoretical interest, with a discrepancy between early practical mathematics and theoretical engineering mathematics (Fig. 1.7).

The European students feel that advanced mathematics is useful, that the role of a teacher is more to help students to apply mathematics than to only transmit knowledge. The Russian students fall in between the two groups and are more diverse in their opinions [32, 33].

Apart from the country and the institution, which do explain a lot of the variability, we looked for characters separating students into groups in a statistically

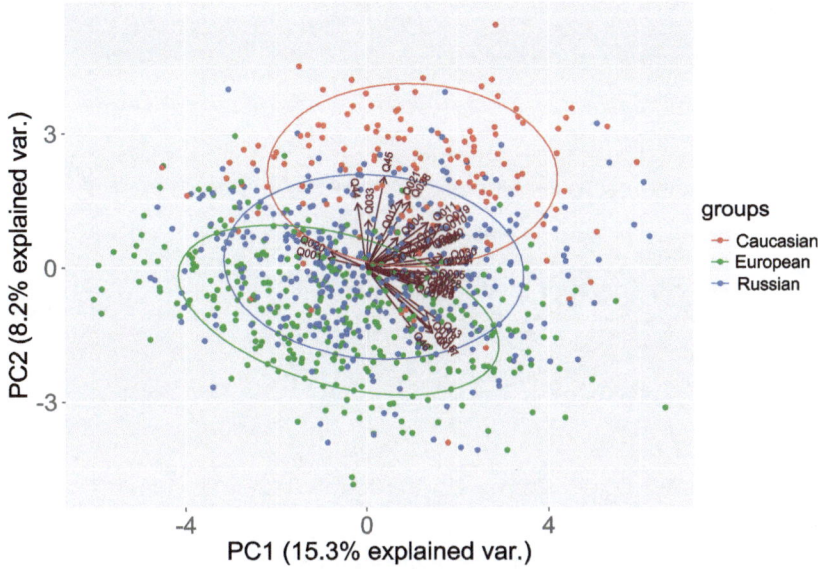

Fig. 1.7 PCA grouped by regions

significant way. In engineering courses, gender is a major differentiating trait [2, 28]. And, to our surprise, the partner's institution explains much better the differences between students than gender: male and female students have very similar responses, only 6 out of 35 questions are statistically distinguishable (p-value $<$ 0.05) and we have no clear-cut semantic explanation of the slight differences: male students tend to disagree a little bit more strongly to the proposal that mathematics can be applied more easily to man crafted objects than to objects found in nature, while female students tend to find slightly more that mathematics courses are practical enough. But the differences are much higher between partner institutions than between genders: there are statistically greater differences between the answers of a student in St Petersburg Electrotechnical University (LETI) and another in Ogarev Mordovia State University (OMSU), both in Russia (much lower p-values, with 16 out of 35 being less than 0.05) than between a male and a female student in each university (Figs. 1.8 and 1.9). And the differences are even higher between institutions belonging to different countries. We have to look at the seventh principal component, which is almost meaningless, in order to get a dimension whose interpretation of the variability relates to gender. The same relative irrelevance with respect to age appears: students' perceptions depend on the year of study, but to an extent much lower than the dependency on the institution. We find these results remarkable.

The main finding of this analysis is that there are indeed great differences between students' responses in partners' higher education institutions, with homogeneous European universities tending to see engineering mathematics as a profes-

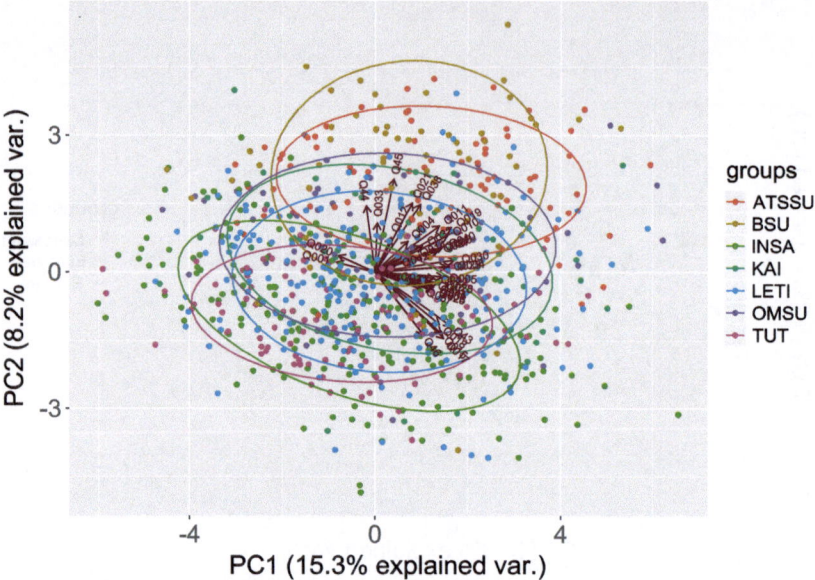

Fig. 1.8 PCA grouped by institutions

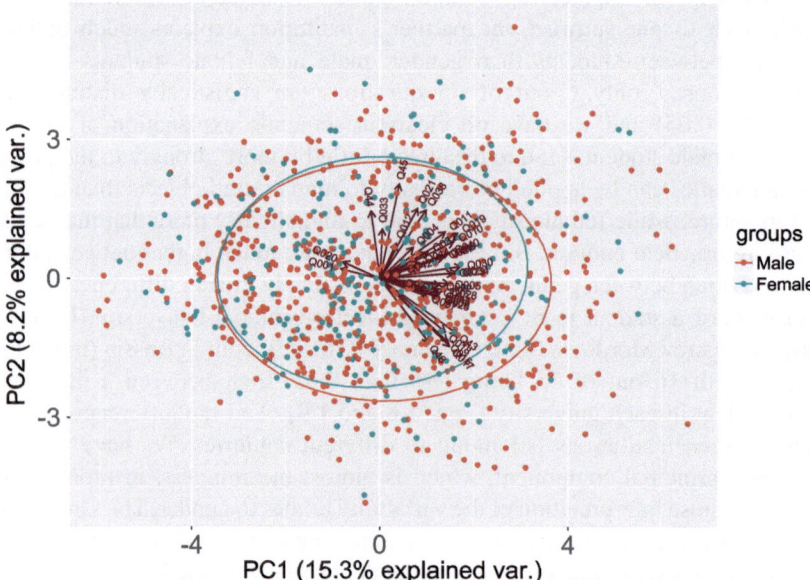

Fig. 1.9 PCA grouped by gender

sional tool on the one side; homogeneous Caucasian universities on the other, where advanced mathematics are felt as dealing with abstraction, and Russian universities in between.

1.3.5   Conclusions

In this study, we observed that European countries on the one hand and South Caucasian countries on the other are quite aligned. However, Russian students' perception is more spread out and in between those of the European and South Caucasian students. The country factor has a large influence but within these differences, institutions can be more finely differentiated and this difference is higher than most other criteria, like gender: a student can be linked to his/er university in a more confident way than to his/her gender or his/her year of study. Comparison with other institutions would be interesting.

The main implication for the MetaMath and MathGeAr projects from this study is that if the European way is to be promoted, the project should put forward the applications of advanced mathematics and focus on competencies rather than on transmission of knowledge.

This questionnaire has some limitations. For instance, its item-internal consistency reliability was not high enough regarding the three dimensions of the questionnaire which we identified a priori. The item-internal consistency reliabilities measured by Cronbach's alpha are 0.52, 0.65, and 0.62, which tells us that reality is more complex than our question choices based on epistemology. It evokes the need to further study to qualify the questionnaire with a bigger homogeneous sample and/or redesign of the current questionnaire by adding more items related to these dimensions or qualify these dimensions better.

Because perceiving mathematics in a positive way would influence students' motivation and performance, it is desirable to change the mathematics contents and the way we teach it in order to address the negative aspects of the perceptions identified here, for instance teaching mathematics as a powerful modelling tool not abstractly, but in actual students' projects.

We might as well try to directly modify students' perceptions by better informing them about some aspects of mathematics; its usefulness in engineer's profession for example. Therefore, we need to know which type of mathematics in-service engineers do use in a conscious way, and what their perceptions are of the mathematics they received during their education.

The current study suggests further investigation avenues: the first one is to study deeper the influence of engineering students' perceptions on mathematics performance for each partner institution. The second one is the elaboration of questionnaires targeting engineers in order to study the perceptions and actual usage of mathematics by professionals. Because the link between students and engineers goes through teachers, we need to study as well the perceptions of teachers themselves. We have already adapted this questionnaire in order to address these two targets and it will be the subject of subsequent articles. This study is only the first real size pilot of a series of further studies to come.

References

1. ACME. (2011). Mathematical needs: the mathematical needs of learners. Report. Advisory Committee on Mathematics Education.
2. Alegria, S. N., Branch, E. H. (2015). Causes and Consequences of Inequality in the STEM: Diversity and its Discontents. International Journal of Gender, Science and Technology, 7(3), pp. 321–342.
3. Angoff, W. H. (1988). Validity: An evolving concept. In H. Wainer, & H. Braun (Eds.) Test validity (pp. 19–32), Hillsdale: Lawrence Erlbaum Associates.
4. Araque, F., Roldán, C., Salguero, A. (2009). Factors influencing university drop out rates. Computers & Education, 53(3), pp. 563–574.
5. Atkins, M., Ebdon, L. (2014). National strategy for access and student success in higher education. London: Department for Business, Innovation and Skills. Accessed 25 October 2016 from https://www.gov.uk/government/uploads/system/uploads/attachment_data/file/299689/bis-14-516-national-strategy-for-access-and-student-success.pdf
6. Benilan, P., Froehly, C. (2002). Les formations supérieures en mathématiques orientées vers les applications - rapport d'évaluation. Comité national d'évaluation des établissements publics à caractère scientifique, culturel et professionnel, France. Accessed 26 October 2016 from http://www.ladocumentationfrancaise.fr/var/storage/rapports-publics/054000192.pdf
7. Bonnay, D., Dubucs, J. (2011). La philosophie des mathématiques. Published online. Accessed 26 October 2016 from https://hal.archives-ouvertes.fr/hal-00617305/file/bonnay_dubucs_philosophie_des_mathematiques.pdf
8. Breiteig, T., Grevholm, B, Kislenko, K. (2005). Beliefs and attitudes in mathematics teaching and learning, In I. Stedøy (Ed.), Vurdering i matematikk, hvorfor og hvor? : fra småskoletil voksenopplæring : nordisk konferanse i matematikkdidaktikk ved NTNU (pp. 129–138), Trondheim: Norwegian University of Science and Technology.
9. Brusilovsky, P., Sosnovsky, S., Yudelson, M. (2009). Addictive links: the motivational value of adaptive link annotation. New Review of Hypermedia and Multimedia, 15(1), pp.97–118.
10. Byrne, M., Flood, B. (2008). Examining the relationships among background variables and academic performance of first year accounting students at an Irish University. Journal of Accounting Education, 26(4), pp. 202–212.
11. Cardoso, W. (1985). Os obstáculos epistemológicos segundo Gaston Bachelard. Revista Brasileira de História da Ciência, 1, pp. 19–27.
12. Chaper, E. A. (2008). The impact of middle school students' perceptions of the classroom learning environment on achievement in mathematics. Doctoral Dissertation. University of Massachusetts Amherst.
13. Charette, Robert N. "An Engineering Career: Only a Young Person's Game?" Institute of Electrical and Electronics Engineers website (Accessed from http://spectrum.ieee.org/riskfactor/computing/it/an-engineering-career-only-a-young-persons-game).
14. Chaudhary, M. P. (2013). Utility of mathematics. International Journal of Mathematical Archive, 4(2), 76–77. Accessed 26 October 2016 from http://www.ijma.info/index.php/ijma/article/view/1913/1133.
15. De Corte, E., Op't Eynde, P. (2002). Unraveling students' belief systems relating to mathematics learning and problem solving. In A. Rogerson (Ed.), Proceeding of the International Conference "The humanistic renaissance in mathematics education", pp. 96–101, Palermo, Italy.

16. Dogan-Dunlap, H. (2004). Changing students' perception of mathematics through an integrated, collaborative, field-based approach to teaching and learning mathematics. Online Submission. Accessed 25 October 2016 from http://files.eric.ed.gov/fulltext/ED490407.pdf.
17. EURYDICE (2011). Mathematics in Europe: Common Challenges and National Policies,. (http://eacea.ec.europa.eu/education/eurydice).
18. The Final Report of the National Mathematics Advisory Panel (2008). U.S. Department of Education.
19. Furinghetti, F., Pehkonen, E. (2002). Rethinking characterizations of beliefs. In G. C. Leder, E. Pehkonen, G. Törner (Eds.), Beliefs: A hidden variable in mathematics education? (pp. 39–57). Springer Netherlands.
20. Githua, B. N. (2013). Secondary school students' perceptions of mathematics formative evaluation and the perceptions' relationship to their motivation to learn the subject by gender in Nairobi and Rift Valley Provinces, Kenya. Asian Journal of Social Sciences and Humanities, 2(1), pp. 174–183.
21. Hanushek E. A., Wößmann L. The Role of Education Quality in Economic Growth. World Bank Policy Research Working Paper No. 4122, February 2007. (http:/info.worldbank.org/etools/docs/library/242798/day1hanushekgrowth.pdf)
22. Henderson S., Broadbridge P. (2009). Engineering Mathematics Education in Australia, MSOR Connections Vol. 9, No 1, February - April.
23. Heublein, U., Schmelzer, R., Sommer, D. (2008). Die Entwicklung der Studienabbruchquote an den deutschen Hochschulen: Ergebnisse einer Berechnung des Studienabbruchs auf der Basis des Absolventenjahrgangs 2006. Projektbericht. HIS Hochschul-Informations-System GmbH. Accessed 12 April 2013 from http://www.his.de/pdf/21/20080505_his-projektbericht-studienabbruch.pdf.
24. Heublein, U., Spangenberg, H., & Sommer, D. (2003). Ursachen des Studienabbruchs: Analyse 2002. HIS, Hochschul-Informations-System.
25. Huikkola M., Silius K., Pohjolainen S. (2008). Clustering and achievement of engineering students based on their attitudes, orientations, motivations and intentions, WSEAS Transactions on Advances in Engineering Education, Issue 5, Volume 5, May, pp. 343–354.
26. Jonassen, D. H. (1995). Supporting Communities of Learners with Technology: A Vision for Integrating Technology with Learning in Schools, Educational Technology, 35, 4, pp. 60–63.
27. Johnson, R. A., Wichern, D. W. (1999). Applied multivariate statistical analysis (6th ed). Upper Saddle River, New Jersey: Prentice-Hall. Accessed 25 October 2016 from http://www1.udel.edu/oiss/pdf/617.pdf
28. Jumadi, A. B., Kanafiah, S. F. H. M. (2013). Perception towards mathematics in gender perspective. In Proceedings of the International Symposium on Mathematical Sciences and Computing Research 2013 (iSMSC 2013) (pp. 153–156). Accessed 25 October 2016 from http://lib.perak.uitm.edu.my/system/publication/proceeding/ismsc2013/ST_09.pdf.
29. Karp, Alexander, Bruce Ramon Vogeli (2011). Russian mathematics education: Programs and practices. Vol. 2. World Scientific.
30. Kohlhase, M. (2006). OMDoc – An open markup format for mathematical documents [version 1.2]. Berlin/Heidelberg, Germany: Springer Verlag.
31. Lassibille, G., Navarro Gómez, L. (2008). Why do higher education students drop out? Evidence from Spain. Education Economics, 16(1), pp. 89–105.
32. Lealdino Filho, P., Mercat C., El-Demerdash, M. (2016). MetaMath and MathGeAr projects: Students' perceptions of mathematics in engineering courses, In E. Nardi, C. Winsløw T. Hausberger (Eds.), Proceedings of the First Conference of the International Network for Didactic Research in University Mathematics (INDRUM 2016) (pp. 527–528). Montpellier, France: University of Montpellier and INDRUM.
33. Lealdino Filho, P., Mercat C., El-Demerdash, M., Trgalová, J. (2016). Students' perceptions of mathematics in engineering courses from partners of MetaMath and MathGeAr projects. 44th SEFI Conference, 12–15 September 2016, Tampere, Finland.

34. Lloyd, G. M., Wilson, M. (1998). Supporting Innovation: The Impact of a Teacher's Conceptions of Functions on His Implementation of a Reform Curriculum. Journal for Research in Mathematics Education, 29(3), pp. 248–274.
35. McLeod, D. B. (1992). Research on affect in mathematics education: A reconceptualization. In D. A. Grouws (Ed). Handbook of research on mathematics teaching and learning (pp. 575–596). New York, NY, England, Macmillan Publishing Co.
36. Melis, E., Andres, E., Büdenbender, J., Frischauf, A., Goguadze, G., Libbrecht, P., Ullrich, C. (2001). ActiveMath: a generic and adaptive Web-based learning environment. International Journal of Artificial Intelligence in Education, 12(4), pp. 385–407.
37. Mullis I., Martin M., Foy P., Arora A., (TIMSS 2011). Trends in International Mathematics and Science Study, International Results in Mathematics. (Accessed from http://timssandpirls.bc.edu/).
38. Mutodi, P., Ngirande, H. (2014). The influence of students' perceptions on mathematics performance. A case of a selected high school in South Africa. Mediterranean Journal of Social Sciences, 5(3), p. 431.
39. Niss, M., Højgaard T., (Eds.) (2011). Competencies and Mathematical Learning, Ideas and inspiration for the development of mathematics teaching and learning in Denmark, English edition, October 2011, Roskilde University, Department of Science, Systems and Models, IMFUFA P.O. Box 260, DK - 4000 Roskilde. (Accessed from http://diggy.ruc.dk/bitstream/1800/7375/1/IMFUFA_485.pdf).
40. NSF Report (2007). Moving forward to improve engineering education #NSB-07-122.
41. OECD (2015). Students, Computers and Learning: Making the Connection, PISA, OECD Publishing. (Accessed from http://dx.doi.org/10.1787/9789264239555-en).
42. OECD (2008). Report on Mathematics in Industry, Global Science Forum, July. (Accessed from https://www.oecd.org/science/sci-tech/41019441.pdf).
43. PISA (2009). Assessment Framework: Key competencies in reading, mathematics and science. (Accessed from http://www.oecd.org/pisa/pisaproducts/44455820.pdf).
44. PISA (2012). The Programme for International Student Assessment. (Accessed from http://www.oecd.org/pisa/keyfindings/pisa-2012-results.htm).
45. Ponte, J. P. (1994). Knowledge, beliefs, and conceptions in mathematics teaching and learning. In L. Bazzini (Ed.), Proceedings of the 5th International Conference on Systematic Cooperation between Theory and Practice in Mathematics Education (pp. 169–177). Pavia: ISDAF.
46. Reusser, K., (1995). From Cognitive Modeling to the Design of Pedagogical Tools. In S. Vosniadou, E. De Corte, R. Glaser, H. Madl (Eds.). International Perspectives on the Design of Technology-Supported Learning Environments, Hillsdale, NJ: Lawrence Erlbaum, pp. 81–103.
47. Richardson, M. (2009). Principal component analysis, Accessed 25 October 2016 at http://www.sdss.jhu.edu/szalay/class/2015/SignalProcPCA.pdf
48. Ruokamo H., Pohjolainen S. (1998). Pedagogical Principles for Evaluation of Hypermedia Based Learning Environments in Mathematics, Journal for Universal Computer Science, 4 (3), pp. 292–307, (Accessed from http://www.jucs.org/jucs_4_3/)
49. SEFI (2013), A Framework for Mathematics Curricula in Engineering Education. (Eds.) Alpers, B., (Assoc. Eds) Demlova M., Fant C-H., Gustafsson T., Lawson D., Mustoe L., Olsson-Lehtonen B., Robinson C., Velichova D. (Accessed from http://www.sefi.be).
50. SEFI (2002). Mathematics for the European Engineer. (Eds.) Mustoe L., Lawson D. (Accessed from http://sefi.htw-aalen.de/Curriculum/sefimarch2002.pdf).
51. Sosnovsky, S., Dietrich, M., Andrés, E., Goguadze, G., Winterstein, S., Libbrecht, P., Siekmann, J., Melis, E. (2014). Math-Bridge: Closing Gaps in European Remedial Mathematics with Technology Enhanced Learning. In T. Wassong, D. Frischemeier, P. R. Fischer, R. Hochmuth, P. Bender (Eds.), Mit Werkzeugen Mathematik und Stochastik lernen - Using Tools for Learning Mathematics and Statistics. Berlin/Heidelberg, Germany, Springer, pp. 437–451
52. Tapia, M., Marsh, G. E. (2004). An Instrument to Measure Mathematics Attitudes. Academic Exchange Quarterly, 8(2), pp. 16–21.

53. Thompson, A. G. (1992). Teachers' beliefs and conceptions: A synthesis of the research. In D. A. Grouws (Ed), Handbook of research on mathematics teaching and learning: A project of the National Council of Teachers of Mathematics (pp. 127–146). New York, NY, England: Macmillan Publishing Co.
54. Tipping, M. E., Bishop, C. M. (1999). Probabilistic principal component analysis. Journal of the Royal Statistical Society: Series B (Statistical Methodology), 61(3), pp. 611–622.

Chapter 2
Methodology for Comparative Analysis of Courses

Sergey Sosnovsky

2.1 Introduction

One of the main goals of the projects MetaMath and MathGeAr was to conduct a range of comparative case studies that would allow the consortium to understand the set and the magnitude of differences and commonalities between the ways mathematics is taught to engineers in Russia, Georgia, Armenia and in the EU. Equipped with this knowledge the consortium was able to produce recommendations for modernisation of existing Russian, Georgian and Armenian mathematics courses during the project.

However, before such comparative analyses can even begin, one needs a clear set of criteria that would determine the nature and the procedure of the planned comparison. Identification of these criteria was the goal of this methodology development. The results of this activity are described in this chapter.

The methodology has been developed by the EU partners (especially, Saarland University and Tampere University of Technology) in consultation with Russian, Georgian and Armenian experts. Dedicated methodology workshops have been organised by the Association for Engineering Education of Russia in the Ministry of Education and Science in Moscow, Russia, and in the National Center for Educational Quality Enhancement in Tbilisi, Georgia, in June, 2014. During these workshops, the first draft of the comparison criteria has been discussed with the partners and invited experts. It has been amended in the future versions of the Methodology to better match the realities of the university education in Russia, Georgia and Armenia and provide a more objective picture of it with regards to mathematics for engineers. Also, during the workshop, the SEFI Framework for

S. Sosnovsky (✉)
Utrecht University, Utrecht, the Netherlands
e-mail: s.a.sosnovsky@uu.nl

© The Author(s) 2018
S. Pohjolainen et al. (eds.), *Modern Mathematics Education for Engineering Curricula in Europe*, https://doi.org/10.1007/978-3-319-71416-5_2

Mathematical Curricula in Engineering Education [1] has been introduced to the Russian, Georgian and Armenian partners as the unified instrument to describe and compare the content of the corresponding courses.

2.2   Criteria for Conducting Comparative Case Studies

2.2.1   University/Program Profile

When comparing courses, it is not enough to choose courses with similar titles. The goals and the purpose of the courses should align and the focus of the courses and their roles within the overall curricula should be comparable. Even the character of the university where a course is offered can make a difference. A classic university and a university of applied sciences can have very different perspectives on what should be the key topics within courses with the same name. In a larger university a professor can have a much richer set of resources than in a smaller one; at the same time, teaching a course to several hundreds of students puts a much bigger strain on a professor than teaching it to several dozens of them. The overall program of studies that the target course is part of is equally important for similar reasons. Therefore, the first set of criteria characterising a course profile focus on the general description of the university and the program (major) where the course is taught. These parameters include:

- Criterion A: University profile
    Classic or applied
    Overall number of students
    Number of STEM disciplines
    Number of STEM students
- Criterion B: Program/discipline profile
    Theoretical or applied
    Number of students
    Role/part of mathematics in the study program

2.2.2   Course Settings

The next set of criteria describes the context of the course including all its organisational settings and characteristics not directly related to pedagogical aspects or the content. This is the metadata of the course, which allows us to easily identify whether the two courses are comparable or not. For example, if in one university a course is taught on a MSc level and in another on a BSc level, such courses are not directly comparable, because the levels of presentation of the course material would differ much. If in one university a course's size is 3 ECTS credits and in the other 7

ECTS, such courses are not the best candidates for comparison either, because the amount of work students need to invest in these two courses will be very different even if the titles of the courses are similar. Sometimes, we had to relax some of these conditions if for particular universities best matches cannot be found. The complete list of course characteristics include:

- Criterion C: Course type
    Bachelor or Master level
    Year/semester of studies (1/2/. . .)
    Selective or mandatory
    Theoretical/applied
- Criterion D: Relations to other courses in the program
    Prerequisite courses
    Outcome courses
    If the course is a part of a group/cluster (from which it can be selected), other courses in the group
- Criterion E: Department teaching the course
    Mathematical/graduating/other
- Criterion F: Course load
    Overall number of credits according to ECTS regulations
    Number of credits associated with particular course activities (lectures/tutorials/practical work/homework/etc.)

2.2.3   Teaching Aspects

In order to describe how the teacher organises the course, we identify three important criteria: use of any particular didactic approach (such as project-based learning, inquiry-based learning, blended learning, etc.), organisation of course assessment (how many tests and exams, what form they take, how they and the rest of the course activity contribute to the final grade) and the resources available to a teacher—from the help of teaching assistants to the availability of computer labs. Teaching aspects include:

- Criterion G: Pedagogy
    Blended learning
    Flipped classroom
    MOOC
    Project-based learning
    Inquiry-based learning
    Collaborative learning
    Game-based learning
- Criterion H: Assessment
    Examinations (how many, oral/written/test-like)
    Testing (how often)

Grade computation (contribution of each course activity to the final grade, availability extra credits)
- Criterion I: Teaching resources
    Teaching hours
    Preparatory hours
    Teaching assistants (grading/tutorials)
    Computer labs

2.2.4 Use of Technology

One of the aims of the MetaMath and MathGeAr projects is to examine and ensure the effective use of modern ICT in math education. Therefore, a dedicated group of criteria has been selected to characterise the level of application of these technologies in the target courses. There are two top-level categories of software that can be used to support math learning: the instruments that help students perform essential math activities and the tools that help them to learn mathematics. The former category includes such products as MATLAB, Maple, Mathematica, or SPSS. These are, essentially, the systems that a professional mathematician, engineer or researcher would use in everyday professional activity. Using them in a course not only helps automate certain computational tasks but also leads to mastering these tools, which is an important mathematical competency on its own. The latter systems are dedicated educational tools. They help students understand mathematical concepts and acquire general mathematical skills. In MetaMath and MathGeAr projects, we apply a particular tool like this—an intelligent education platform Math-Bridge. In both these categories, the number and diversity of available systems is very large. The focus of the criteria in this set is to detect whether any of these systems are used and to what degree, namely what the role in the course is.

- Criterion J: Use of math tools
    Name of the tool(s) used (MATLAB, Maple, MathCAD, Mathematica, SPSS, R, etc.)
    Supported activities (tutorials, homework)
    Overall role of the tool (essential instrument that must be learnt or one way to help learn the rest of the material)
- Criterion K: Use of technology enhanced learning (TEL)-systems
    Name and type of tool used (Geogebra—math simulation; STACK—assessment software; Math-Bridge—adaptive learning platform; etc.)
    Supported activity (assessment, homework, exam preparation)
    Role on the course (mandatory component/extra credit opportunity/fully optional supplementary tool)

2.2.5 Course Statistics

Another important aspect of the course is the data collected about it over the years. It shows the historic perspective and evolution of the course, and it can also provide some insights into the course's difficulty and the profile of a typical student taking a course. Although by itself this information might be not as important for course comparison, combined with other criteria it can provide important insights.

- Criterion L: Course statistics
  Average number of students enrolled in the course
  Average percentage of students successfully finishing the course
  Average grade distribution
  Percentage of international students
  Overall student demographics (gender, age, nationality, scholarships, etc.)
  Average rating of the course by students

2.2.6 Course Contents

Finally, the most important criterion is the description of the learning material taught in the course. In order to describe the content of the analysed courses in a unified manner that would allow for meaningful comparison we needed a common frame of reference. As the context of mathematical education in this project is set for engineering and technical disciplines, we have decided to adopt a "Framework for Mathematics Curricula in Engineering Education" prepared by the Mathematics Working Group of the European Society for Engineering Education (SEFI) [1]. This report is written about every 10 years; and the current edition formalises the entire scope of math knowledge taught to engineering students in EU universities in terms of competences. The competencies are broken into four levels, from easier to more advanced, and allow for composite representation of any math course. As a result, every course can have its content described in terms of atomic competencies and two similar courses can easily be compared based on such descriptions.

- Criterion M: Course SEFI competency profile
  Outcome competencies of the course (what a student must learn in the course)
  Prerequisite competencies of the course (what a student must know before taking the course)

2.3   Application of the Criteria for the Course Selection and Comparison

This set of criteria should be used (1) for selecting appropriate courses for the comparison and (2) for conducting the comparison itself. During the selection process, Criteria A and B will ensure that only universities and study programs with matching profiles are selected. Criteria C, D, E and F will help to select the courses that correspond in terms of their metadata. Criteria G, H and I will help to filter out courses that utilise unconventional pedagogical approaches or differ too much in terms of assessment organisation and teaching resources available.

At this point, if a pair of courses passed the screening, they can be safely compared; all criteria starting J contribute to the comparison. One needs to note that, in some cases, the strict rules of course selection might not apply, as a particular partner university sometimes presents a very unique case. In such situations, the selection rules can be relaxed.

Reference

1. SEFI (2013). A Framework for Mathematics Curricula in Engineering Education. (Eds.) Alpers, B., (Assoc. Eds) Demlova M., Fant C-H., Gustafsson T., Lawson D., Mustoe L., Olsson-Lehtonen B., Robinson C., Velichova D. (http://www.sefi.be).

Chapter 3
Overview of Engineering Mathematics Education for STEM in Russia

Yury Pokholkov, Kseniya Zaitseva (Tolkacheva), Mikhail Kuprianov, Iurii Baskakov, Sergei Pozdniakov, Sergey Ivanov, Anton Chukhnov, Andrey Kolpakov, Ilya Posov, Sergey Rybin, Vasiliy Akimushkin, Aleksei Syromiasov, Ilia Soldatenko, Irina Zakharova, and Alexander Yazenin

Y. Pokholkov · K. Zaitseva (Tolkacheva)
Association for Engineering Education of Russia (AEER), Tomsk, Russia
e-mail: pyuori@mail.ru

M. Kuprianov · I. Baskakov · S. Pozdniakov · S. Ivanov (✉) · A. Chukhnov · A. Kolpakov ·
V. Akimushkin
Saint Petersburg State Electrotechnical University (LETI), St. Petersburg, Russia
e-mail: mskupriyanov@mail.ru; bosk@bk.ru; sg_ivanov@mail.ru

I. Posov
Saint Petersburg State Electrotechnical University (LETI), St. Petersburg, Russia

Saint Petersburg State University (SPbU), St. Petersburg, Russia
e-mail: i.posov@spbu.ru

S. Rybin
Saint Petersburg State Electrotechnical University (LETI), St. Petersburg, Russia

ITMO University, Department of Speech Information Systems, St. Petersburg, Russia

A. Syromiasov
Ogarev Mordovia State University (OMSU), Department of Applied Mathematics,
Differential Equations and Theoretical Mechanics, Saransk, Russia
e-mail: syal1@yandex.ru

I. Soldatenko · A. Yazenin
Tver State University (TSU), Information Technologies Department, Applied Mathematics
and Cybernetics Faculty, Tver, Russia
e-mail: soldis@yandex.ru; Yazenin.AV@tversu.ru

I. Zakharova
Tver State University (TSU), Mathematical Statistics and System Analysis Department,
Applied Mathematics and Cybernetics Faculty, Tver, Russia
e-mail: zakhar_iv@mail.ru

© The Author(s) 2018
S. Pohjolainen et al. (eds.), *Modern Mathematics Education for Engineering
Curricula in Europe*, https://doi.org/10.1007/978-3-319-71416-5_3

3.1 Review of Engineer Training Levels and Academic Degrees

Higher education in general and engineering education in particular are divided into three levels in Russia. The first level is the Bachelor's degree course; the second level is the specialist and Master's degree program; the third level is the postgraduate training program.

"The Bachelor's degree" training level was introduced in Russia in 1996. The standard training period to get the Bachelor qualification (degree) is at least 4 years with the total study load volume of 240 credits. The Bachelor qualification is conferred based on the results of a presentation of the graduate thesis at a session of the State Certifying Commission.

In 1993 the term of Master was established in Russia as the qualification of the graduates of educational institutions of higher professional education. The standard period of the Master training program (for the intramural form of study) is currently 2 years with the credit value of the educational program of 120 credits. Before that, however, the student is to complete the Bachelor (4 years) or specialist (5–6 years) training program. The Master's qualification (academic degree) is conferred only after presentation of the Master's thesis at a session of the State Certifying Commission.

The Bachelor's and Master's qualifications (degrees) were historically preceded by the specialist qualification presupposing 5–6 years of continuous learning. In the Soviet Union it was the only possible qualification; but then the gradual transition to the Bachelor and Master levels took place. At present the specialist qualification has been preserved. When a prospective student applies documents to a university he or she may be admitted to a Bachelor or specialist's training program (depending on the selected department, future profession, etc.). Today, however, the specialist qualification is comparatively rare, having receded in favor of the Bachelor's training program. It is conferred based on the results of the presentation of a graduation project or graduation thesis at the session of the State Certifying Commission and gives the right to enter the Master's Degree course (although like the Master's Degree course the specialist degree course is the second level of higher education) and the postgraduate training program.

The postgraduate program is one of the forms of training of top-qualification personnel. Before September 1, 2013 the postgraduate program was one of major forms of training of the academic and scientific personnel in the system of postgraduate professional education. Since September 1, 2013 (the date when the Federal Law No. 273-FZ dated December 29, 2012 "On Education in the Russian Federation" came into force) the postgraduate program was referred to the third level of higher education. The person who has completed the postgraduate program and presented a thesis receives the academic degree of a candidate of the sciences. In the USSR, the Russian Federation (RF) and in a number of Commonwealth of Independent States (CIS) countries this degree corresponds to the Doctor of Philosophy degree (PhD) in western countries. Presentation of a candidate thesis is public and takes place in a special dissertation council in one or several related scientific specialties. In many cases the thesis is presented outside of the higher educational institution where the applicant for the degree studied.

A Doctor of Sciences is the top academic degree (after the Candidate of Sciences). In Russia the degree of a Doctor of Sciences is conferred by the Presidium of the Higher Attestation Commission (VAK) of the Ministry of Education and Science of the Russian Federation based on the results of the public presentation of the doctorate thesis in a specialized dissertation council. The applicant for the degree of a Doctor of Sciences is to have the academic degree of a Candidate of Sciences. An approximate analog of the Russian doctoral degree accepted in Anglo-Saxon countries is the degree of a Doctor of Sciences (Dr. Sc.) or the German Doctor habilitatis degree (Dr. habil.).

3.2 Forms of Studies of Engineering Students

Three forms of obtaining education have been traditionally established in Russia: intramural (full-time), part-time (evening time), extramural.

In the intramural form of education the student attends lectures and practical classes every day or almost every day (depending on the timetable). Most classes are held in the morning or in the afternoon (hence the second name).

The evening time form of studies is primarily designed for those students who work during the day. In this connection classes are held in the evening hours. Accordingly, less time is provided for classroom studies and the volume of unsupervised activities increases.

The extramural form of studies presupposes that students meet teachers only during the examination periods which take place 2–3 times a year and are 1–2 weeks long each. During these periods students have classroom studies and take tests and exams; besides, students get assignments which should be done in writing by the beginning of the next examination period. Thus, in the extramural form of studies the contact work with the teacher is minimal and the volume of unsupervised work is maximum.

The new law of education (to be further discussed below) includes new forms of organization of education: on-line learning and remote learning. Network learning presupposes that a student studies some subjects in one higher educational institution and other subjects in another and then gets a "combined" diploma of both institutions. In remote learning the students communicates with the teacher mostly by means of Internet and classes are conducted in the form of a video conference.

3.3   Statistics and Major Problems of Engineering Education in Russia

The Russian Federation has 274 engineering higher educational institutions, training 1,074,358 students. With account of comprehensive universities also admitting students for engineering training programs the total number of higher educational institutions where a student can obtain engineering education is 560. The total number of engineering students is about a million and a half. The distribution of engineering students by regions of Russia[1] is shown on the map in Fig. 3.1. The numbers on the map are described in Table 3.1.

The number of students per 10 thousand inhabitants of the population varies from 78 (East-Siberian region) to 295 (Northwest region) but in other regions the distribution is more uniform, and it varies between 150 and 200 students per 10,000 inhabitants.

The problems of Russian engineering education include:

- Disproportionality between the distribution of higher educational institutions by regions of Russia and the territorial distribution of production facilities.
- Low quality of admission (weak school knowledge of many prospective students).
- Low level of Russian domestic academic mobility.
- Seclusion from international educational networks.

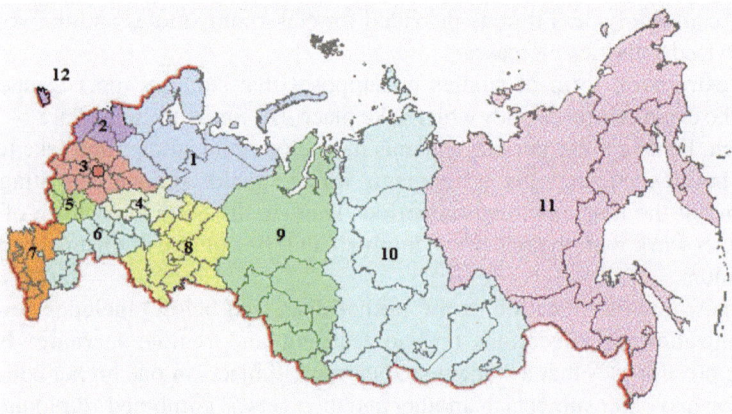

Fig. 3.1 Geographical overview of Russian engineering education

---

[1]Statistics (2016), http://aeer.cctpu.edu.ru/winn/ingobr/tvuz_main.htm.

Table 3.1 Statistics of Russian engineering education

#	Region	Number of engineering students in higher educational institutions	Number of engineering students
1	North region	12	25, 481
2	Northwest region	24	123, 386
3	Central region	92	338, 002
4	Volga-Vyatka region	7	31, 455
5	Central Black Earth region	10	38, 819
6	Volga region	34	114, 324
7	North Caucasus region	21	81, 650
8	Ural region	23	103, 071
9	West-Siberian region	24	119, 207
10	East-Siberian region	20	66, 462
11	Far East region	6	28, 039
12	Kaliningrad region	1	4462
	Total	274	1, 074, 358

3.4 Regulatory Documents

The system of state standardization of higher education program, acting from the mid-1990s, since the introduction of federal state standards (FSES 3), is relaxing strict regulation of the contents of education in the form of a specified set of subjects with a fixed amount of credits (state educational standards SES-1, SES-2), is now developing towards regulation of the structures of educational programs, conditions of implementation and results of learning (FSES 3, FSES 3+, in the long term FSES 4). For example SES-2 contained a cyclic structure:

- GSE cycle—general humanities and social-economic subjects;
- EN cycle—general mathematical and science subjects;
- OPD—general professional subjects;
- DS—specialization subjects;
- FTD—optional subjects.

The central place in SES-2 is taken by section 4 "Requirements to the compulsory minimal contents of the basic educational training program". This section includes a list of compulsory subjects for every cycle, their credit values in academic hours and a mandatory set of didactic units.

Convergence of the national systems of education within the frames of the European Union is an important landmark in the global development of the higher school in the twenty-first century. The official date of the beginning of the convergence and harmonization process in higher education of European countries with a view to creation of harmonized higher education is considered to be June 19, 1998 when the Bologna Declaration was signed. Russia joined the Bologna process in 2003. As a result of this joining the educational process in most European countries is currently in the process of reforming. Higher educational institutions have set the task of not to unify but to harmonize their educational programs with others. In this connection, the state educational standards are undergoing rethinking and considerable changes.

With the introduction of the federal state standards of the third generation (FSES 3, 2011) Russian higher educational institutions gain greater independence in the formation of the major educational programs, choice of the learning contents, forms and methods, which enables them to compete on the market of educational services, and to respond to the demands of the labor market.

One of the major distinguishing features of the new standards is the competency-based approach. The essence of this approach is that the focus of the educational process is transferred from the contents of education to the results of studies which should be transparent, i.e. clear to all the stakeholders: employers, teachers, and students. The results of training are described by means of a system of competencies being a dynamic combination of knowledge, aptitudes, skills, abilities and personal qualities, which the student can demonstrate after completion of the educational program. The federal state standards of the third generation have inherited a cyclic structure. A major specific feature of the FSES of the higher professional education was the use of credits as a measure of the credit value of educational programs. The indicators of the credit value of educational programs, general speaking, the credit value of the cycles of subjects, are set in educational standards in credit units. For example, the aggregate credit value of the bachelor training is set at 240 credit units, Master training 120 credit units, specialist 300 credit units.

Just as "an academic hour", a "credit unit" is a unit of measurement of the credit value of academic work, but much more consistently oriented towards the work of the student rather than to the teacher. In all international and national systems there is a correspondence between credits units and hours. The method recommended by the Ministry of Education of Russia in 2002 establishes the equivalent of 1 credit unit to correspond to 36 academic hours.

The central place of these standards was taken by the section with a list of study cycles, mandatory subjects for every cycle which regulated the credit value of every cycle in credit units and code of competencies formed in studying the subjects.

Another specific feature of FSES 3+ is the introduction of the postgraduate program (postgraduate military course), residency training and assistantship–traineeship into the levels of higher education.

Pursuant to 273-FZ, dated 29.12.2012, in developing the main curriculum an educational organization independently determines the distribution of the learning material by subjects and modules and establishes the sequence of their study.

The competency-based approach demanded comprehensive restructuring and modernization of the existing education system. Effective use of the competency-based approach is unthinkable without an adequate system of appraisal of every formed competency of the student determined by the state standard as mandatory for the particular educational program. Accordingly, there is need for development and introduction of the fund of means of appraisal allowing such an appraisal to provide a qualified conclusion regarding the conformity of the educational process to regulatory requirements. The need of the availability of such a fund with every educational organization is unequivocally enshrined in the Order of the Ministry of Education and Science of the Russian Federation dated 19.12.2013 No.1367 (revised on 15.01.2015): "20. Appraisal means are presented in the form of the fund of appraisal means for midterm assessment of the learners and for final (state final) assessment. 21. The fund of appraisal means for midterm assessment of learners in the subject (module) or practice included, respectively, into the steering program of the subject (module) or program of practice contains a list of competencies stating the stages of their formation in the process of study of the educational program; description of the indicators and criteria of appraising the competencies at different stages of their formation; description of the appraisal scales, standard assignments for submission or other materials necessary to appraise the knowledge, aptitudes, skills and (or) experience of activities characterizing the stages of formation of the competencies in the process of study of the educational program; guidance materials determining the procedures of appraisal of the knowledge, skills and (or) experience of activities characterizing the stages of formation of the competencies. For every result of study in a subject (module) or practice the organization determines the indicators and criteria of appraising the formedness of competencies at different stages of their formation and the appraisal procedures." The list of universal competencies has been approved by the Ministry of Education and Science of the RF. Universal competencies within the frames of the concept of modern education form the level of development of a specialist distinguishing a specialist with higher education from a specialist of a lower level.

Besides, the methodological recommendations of the Ministry of Education and Science of the Russian Federation dated 22.01.2015 gear educational organizations to taking account of the requirements of the relevant professional standards in the creation of the basic educational programs.

Initially, FSES 3+ was to point to conformity to professional standards. The professional standard of a characteristic of the qualification necessary for an employee to perform a particular kind of professional activities. A professional standard is actually a document containing requirements to:

- the level of the employee's qualification;
- the experience of practical activities, education and learning;
- the contents and quality of the activities;
- the conditions of performance of the labor activities.

As of the moment of issue of FSES 3+ the professional standards in most fields of professional activities had not yet been approved; therefore, FSES 3+ could not formulate the graduate's professional competencies oriented towards generalized labor functions (kinds of professional activities) set by a concrete professional standard (PS). Analysis of the structure of the already approved PS has shown the impossibility to establish a mutually equivocal correspondence between fields of professional activities and educational fields. Therefore, "the core" of training has been identified in FSES 3+ in the form of universal (general culture) competencies and general professional competencies (independent of the particular kind of professional activities for which the learner is preparing and the focus (specialization) of the program). "The core" of training determines the "basic" part of the educational program which is quite fundamental and unalterable. "The variative part" of the program is oriented towards particular generalized labor functions or kind (kinds) of professional activities set by professional standards (if available). This part of the program is to be easily renewable and adaptable to new demands of the labor market. Figure 3.2 presents the structure of the list of education areas in the RF stating the number of consolidated groups of specialties and specializations included in every area of education and the number of such specializations.

Higher educational institutions are currently facing a crucial task: development of educational programs with account of the available professional standards, creation of adequate funds of means of appraisal. Of interest in this connection will be the available experience of the leading Russian universities in this area gained in the implementation of the international project 543851-TEMPUS-1-2013-1-DE-TEMPUS-JPCR (MetaMath) "Modern educational technologies in development of the curriculum of mathematical subjects of engineering education in Russia" and the Russian scientific-methodological projects "Scientific-methodological support

Basis - harmonization with European approaches (FOS 2007, MSCO 2011)	Area of education	CGSS	ATS
	Mathematical and natural sciences	6	50
	Engineering, technologies and technical sciences	23	245
	Healthcare and medical sciences	5	113
Introduction of postgraduate program (postgraduate military course), residency training and assistanceship-traineeship into levels of higher education	Agriculture and agricultural sciences	2	30
	Social sciences	7	65
	Education and pedagogical sciences	1	11
	Art and culture	6	94
	Humanities	5	34
	Defense and security of the state. Military sciences	2	21

Fig. 3.2 List structure of education areas

of development of exemplary basic professional educational programs (EBPEP) by areas of education", "Development of models of harmonization of professional standards and FSES of higher education by fields of study/specialties in the field of mathematical and natural sciences, agriculture and agricultural sciences, social sciences, humanities and levels of education (Bachelor's, Master's, specialist degree programs)". Within the frames of grants working groups of Russian higher educational institutions developed exemplary educational programs of higher education in the modular format under the conditions of the introduction of "framework" federal state educational standards—FSES 3+ and in the long term FSES-4. The developers give practical recommendations for implementation of the competency-based approach in designing and implementing the educational programs.

3.5   Comparison of Russian and Western Engineering Education

Given below is the comparison of Russian and western systems of education in terms of several formal features (such as the number of academic hours allotted for study of the Bachelor's degree program). This analysis certainly cannot be called complete; the contents of the educational program and the teaching quality depend on the particular higher educational institution and department. Nevertheless, such a comparison is to emphasize the greatest similarities and differences between educational systems in Russia and Europe.

The information as regards the structure of domestic educational programs is presented by the example of the curriculum of the field of Software Engineering of Ogarev Mordovia State University (OMSU) as one of the typical representatives of the Russian system of education. An example of a European technical university is Tampere University of Technology, TUT (Finland). The information has been taken from the TUT Study Guide; the authors were oriented towards the degrees of a Bachelor of Science in Technology and a Master of Science (Technology) in the field of Information technology (Tables 3.2, 3.3, 3.4, 3.5, 3.6, 3.7 and 3.8).

Table 3.2   Academic hours and credit units

Russia (OMSU)	European Union (TUT)
1 academic hour = 45 min	1 academic hour = 45 min
1 credit unit (CU) = 36 h (adapted to a 18-week term, weekly credit amount is 1.5 CU)	1 credit unit (credit, cu, ECTS) = 26 2/3 h in Finland (25–30 h in different countries of the EU)
CU includes all kinds of student's work, including independent studies	ECTS includes all kinds of student's work, including independent studies
Credit value of 1 year 60 CU	Credit value of 1 year 60 ECTS
The credit value of a subject is a whole number (at least half-integer) of CU, i.e. a multiple of 36 or 18 h	Formally the credit value of a subject is a whole number of ECTS, although it may be a rounded off number (Discrete Mathematics): 4 ECTS = 105 h)

Table 3.3 Freedom in choosing subjects

Russia (OMSU)	European Union (TUT)
When entering the university the student chooses both the department and the field of study. Thereby the student invariably chooses the greater part of subjects: both in terms of quantity and volume of hours the selected subjects make at least 1/3 of the volume of the variative part. On average this is about 1/6 of the total volume of academic hours (the mandatory and variative parts are approximately equal in terms of volume). Choosing the educational program specialization the student automatically makes the decision on all positions of the professional subjects as chosen	When entering the university the student chooses the study program. Orientation is organized for first-year students before the first week of studies. The student must (by means of a special online-instrument) draw up a Personal Study Plan approved by the department. Most subjects are mandatory but many courses can be freely chosen; the order of their study is recommended but can be modified by the student. The student enrolls for the courses to get the necessary amount of credits (180 for the Bachelor's degree program, 120 more for the Master's degree). Some courses or minors can be taken from another higher educational institution (these may be the so-called minor studies—see below). The students have to annually confirm their plans to continue their education at the TUT (see below)

Table 3.4 Organization of the educational process

Russia (OMSU)	European Union (TUT)
There are lectures and practical classes stipulated in all subjects. The total share of lectures in every cycle of the subjects (GSE, EN, OPD + DS + FTD) is not more than 50%	Courses may contain only lectures or also contain practical exercises, laboratory work or work in groups. These exercises sometimes are mandatory
The details of organization of the educational process in every subject are given at one of the first classes	The details of organization of the educational process in every subject are given at the introductory lecture
Lectures are delivered to the whole class (students of one specialization of training); practical studies are conducted with students of one group	A lecture is delivered to all students that have enrolled for the course. This also refers to practical classes (if they are provided)
There is a possibility to be a non-attending student	There is a possibility to be a non-attending student
A well-performing student is promoted to the next course automatically	A student must annually confirm the desire to continue education
An academic year starts on 1 September and is divided into two terms (autumn and spring). At the end of every term there is an examination period. The "net" duration of a term is 18 weeks	An academic year can start in late August early September and is divided into 2 terms (autumn and spring); every term is divided into two periods (8–9 weeks each). The "net" duration of a term is 17–19 weeks

Table 3.5 Grading system

Russia (OMSU)	European Union (TUT)
Rating system (max 100 points). 70% of the grade are gained during the term. The aggregate rating is converted into the final grade	The rating system is not introduced at the university as a whole; the current and midterm performance is not always taken into account (see above: there may be no practical classes but sometimes they are stipulated, just as mandatory exercises)
A 4-point scale is applied: • Excellent: not less than 86 points out of 100. • Good: 71–85.9 points out of 100. • Satisfactory: 51–70.9 points out of 100. • Unsatisfactory: not more than 50.9 points out of 100. Some Russian higher educational institutions (primarily in Moscow, for example, MEPI) introduced the ECTS grade system in their institutions tying it to the student's rating. The specific weight of the term may differ from 70% accepted at OMSU (for example, 50% at KVFU)	6-point (from 0 to 5) scale is consistent with the European ECTS: • Excellent (ECTS—A): 5 points • Very good (ECTS—B): 4 points • Good (ECTS—C): 3 points • Highly satisfactory (ECTS—D): 2 points • Satisfactory (ECTS—E): 1 point • Unsatisfactory (ECTS—F): 0 points
There are subjects in which "pass" and "fail" grades are given	There are subjects in which "pass" and "fail" grades are given
If there are no grades below "good", with at least 75% "excellent" grades and an "excellent" grade for the FSA a "red" diploma (with honors) is issued. Tests (all passed) are not taken into account in calculation of the grades	If the weighted grade average is not below 4 and the Master's thesis is passed with the grade not below 4, graduation with distinction is issued. Tests are not taken into account in the calculation of the average grade point

Table 3.6 Organization of final and midterm assessment

Russia (OMSU)	European Union (TUT)
Exams	
The timetable of exams is drawn up for a group. The entire group takes the exam on the same day	Exams may be taken at the end of every academic period. Students enter for an exam individually
After a subject has been studied, the group is to take an exam in it automatically	In the case of desire to take an exam, the student is to enter for it in advance (at least a week before it is to be taken)
The "net" duration of an exam for every student is generally about 1 h	The "net" duration of an exam for every student is generally about 3 h
An exam is generally an oral answer	An exam is a written work made on special forms
The grade for an exam is determined right after it is taken	A teacher has a month to check the examination papers
An exam is generally administered by the teacher delivering lectures in the subject (maybe together with an assistant conducting practical classes). He is present at the exam and gives the grade	The exam is conducted by a special employee of the university (invigilator); the teacher delivering the course is not present. But all the remarks made to the student during the exam are recorded by the invigilator on his forms
Final assessment	
The FSA (final state assessment) consists of a state exam (at the option of the university) and a graduation qualification paper	In addition to Bachelor's thesis an exam in the specialization (matriculation exam) is taken and thesis is presented in a seminar. Bachelor's thesis may be carried out as group work; in this case it is necessary to state the contribution of every student in the performance of the assignment
In the Master's degree program the graduation work is prepared in the form of a Master's thesis The final state assessment in the Master's degree program may include a state exam	In addition to a Master's thesis a matriculation exam in the specialization is taken and participation in a Master's seminar
A foreign language is included in the mandatory part of the Bachelor's and Master's degree educational program	Foreign language is included as mandatory part of Bachelor's degree

Table 3.7 Bachelor's degree program

Russia (OMSU)	European Union (TUT)
Period of studies 4 years	Period of studies 3 years (3–4 years in EU countries)
Total credit value of the main curriculum = 240 CU = 8640 h	Total credit value of the main curriculum = 180 ECTS = 4800 h
B.1 GSE cycle: 35–44 CU, with the basic part of 17–22 CU MSU: 38 CU, with the basic part of 20 CU B.2 MiEN cycle: 70–75 CU, with the basic part of 35–37 CU MSU: 75 CU, with the basic part of 36 CU B.3 Prof cycle: 100–105 CU, with the basic part of 50–52 CU MSU: 105 CU, with the basic part of 52 CU B.4 Physical education: 2 CU B.5 Practical training and on-the-job training: 12–15 CU MSU: 12 CU FSA: 6–9 CU	The core studies (basic or central subjects) are the mathematical and natural science subjects as well as other basic courses. The objective is to familiarize the student with basic notions in his field, to give the necessary knowledge for further studies: 90–100 ECTS. Pre-major studies (introduction to the specialization): not more than 20 ECTS Major subject studies are the subject determining the future qualification, including the Bachelor's thesis: 20–30 ECTS Minor subjects (other subjects) are additional subjects but consistent with the future Bachelor's qualification: 20–30 ECTS. Elective studies are not mandatory if the student has fulfilled the minimal requirements of admission for the study program. Practical training is by the decision of the department, not more than 8 ECTS. Bachelor's thesis is an analog of the FSA: 8 ECTS
There are subjects having no direct influence on the future professional skills: the GSE cycle, physical education. The university is considered to train both a specialist and a cultured person	There are no humanities and physical education. The university trains a specialist only; all education is subordinate to this objective. Matriculation examination can be written in Finnish, Swedish or English
Most subjects are fixed in the curriculum	Most subjects are chosen by the student one way or another
Practical training is a mandatory part of the program	The decision about the need for practical training is made by the department
The volume of hours just for training of a specialist (without GSE and physical education): 200 CU = 7200 h	All hours are allocated for training of a specialist, their volume is 180 cu = 4800 h
The volume of the MiEN cycle: 2520–2700 h	The volume of Core studies (analog of the MiEN): 2400–2666.7 h
The volume of the Prof cycle: 3600–3780 h	The volume of Pre-Major + Major + Minor (analog of Prof): 1066.7–2133.3 h

Conclusion: Despite the availability of humanities the Russian Bachelor's degree program contains the same volume of basic knowledge (comparison of MiEN and Core studies) but substantially outstrips the Finnish by the volume of professional training

Table 3.8 Master's degree program

Russia (OMSU)	European Union (TUT)
Period of studies 2 years	Period of studies 2 years
Total credit value of the curriculum = 120 CU = 4320 h	Total credit value of the curriculum = 120 ECTS = 3200 h
M.1 General science cycle: 23–26 CU, with the basic part of 7–8 CU M.2 Professional cycle: 33–36 CU, with the basic part of 10–12 CU M.3 Practical training and research: 48–50 CU M.4 FSA: 12 CU	Common Core studies: 15 ECTS, including compulsory 7 ECTS, complementary 8 ECTS. There are several complementary subjects to choose from; it is just necessary to get 8 ECTS. Major study (basic specialization): 30 ECTS (in this case several variants to choose from) Minor study (other subjects close to the specialization): 25 ECTS (in this case without any choice). Elective studies: 20 ECTS (this includes studies which cannot be included in other section, for example, the English language). Master's thesis—analog to the FSA: 30 ECTS
There are "general" subjects. Their volume (M.1 cycle): 828–936 h	There are "general" subjects. Their volume (common core + elective studies): 933.3 h
Volume of the subjects of the prof. cycle: 1188–1296 h	Volume of the subjects of the prof. cycle (major + minor): 1466.7 h
Volume of Research work + FSA (M.3 + M.4): 2160–2232 h	Volume of research work + FSA (Master's thesis): 800 h

Conclusion: In the EU a Master's degree program more time is spent on training than in Russia (possibly because the volume of the Bachelor's degree program is bigger in Russia and Finnish Masters are still to be educated to the level of the Russian Bachelor's degree level). The research component in the degree is much stronger in Russia

Chapter 4
Overview of Engineering Mathematics Education for STEM in Georgia

David Natroshvili

4.1 Introduction

The three-cycle higher education (HE) system has been introduced in Georgia in 2005, when Georgia became a member of the Bologna Process at the Bergen Summit. Bachelor, Master and Doctoral programs have already been introduced in all stately recognised higher education institutions (HEIs), as well as ECTS and Diploma Supplement. All students below doctoral level are enrolled in a two-cycle degree system (except for certain specific disciplines such as medicine and dental medicine education)—one cycle education and with its learning outcomes corresponding to the Master's level.

There are three types of higher education institutions; see Table 4.1 and Fig. 4.1.

University—a higher education institution implementing the educational programmes of all three cycles of higher education and scientific research;

Teaching University—a higher education institution implementing a higher education programme/programmes (except for Doctoral programmes). A Teaching University necessarily implements the second cycle—the Master's educational programme/programmes;

College—a higher education institution, implementing only the first cycle academic higher education programmes.

HEIs can be publicly or privately founded, but the quality assurance criteria are the same despite the legal status of the institution.

Bachelor's Programme is a first cycle of higher education, which lasts for 4 years and counts 240 ECTS; after completion of this programme students are awarded the Bachelor's Degree (Diploma).

Master's Programme is a second cycle of higher education, which lasts for 2 years with 120 ECTS; after completion of the program students are awarded the

D. Natroshvili (✉)
Georgian Technical University (GTU), Department of Mathematics, Tbilisi, Georgia

© The Author(s) 2018
S. Pohjolainen et al. (eds.), *Modern Mathematics Education for Engineering Curricula in Europe*, https://doi.org/10.1007/978-3-319-71416-5_4

Table 4.1 Table of higher
education institutions in
Georgia

HEI	State	Private	Total
University	12	16	28
Teaching university	7	24	31
College	1	13	13
Total	20	53	73

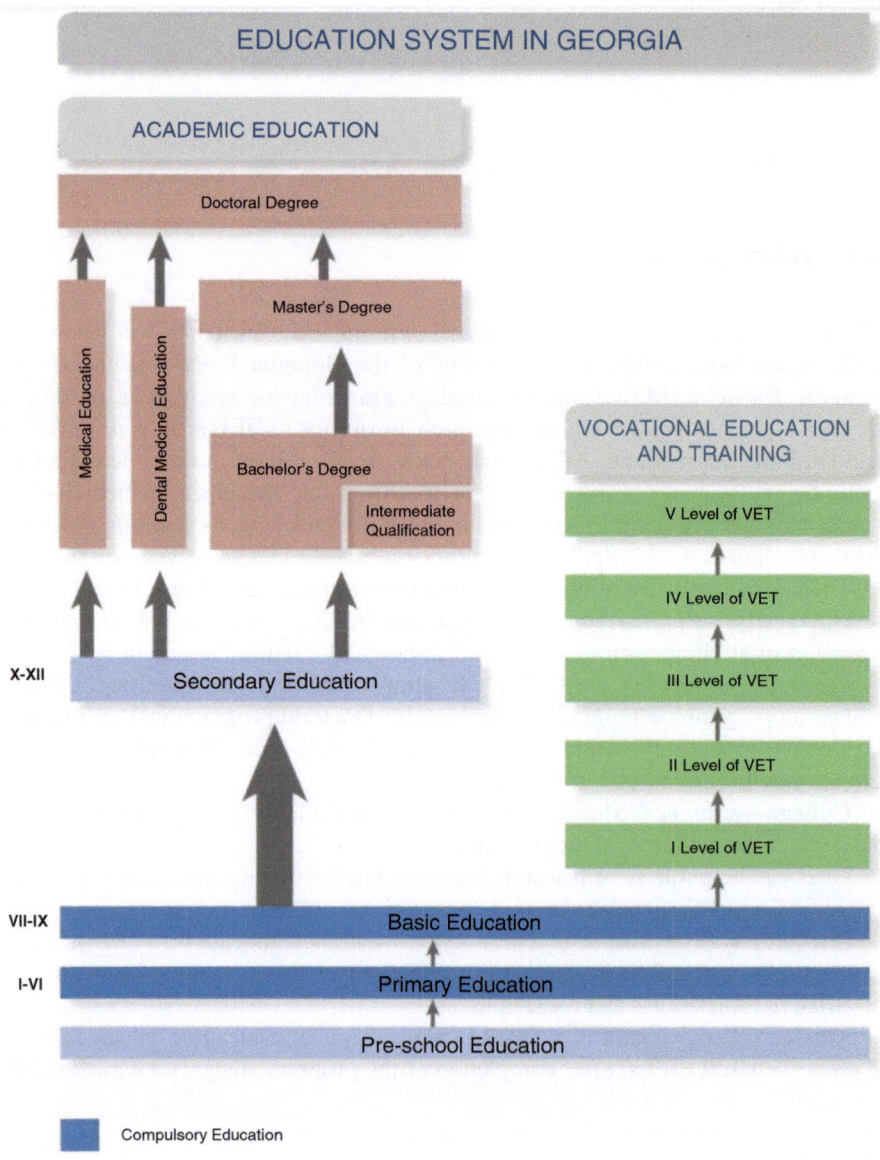

Fig. 4.1 The Georgian education system

Master's Degree (Diploma). Students with Bachelor's Degree Diplomas are required to pass Unified Master's Examinations. The Doctoral Programme is a third cycle of higher education with a minimum duration of 3 years with 180 ECTS; after completion of this programme students are awarded the Doctor's Degree Diplomas. The precondition of entering the third cycle is the completion of the second cycle.

Medical education covers 6 years of studies and counts 360 ECTS.

Dental Medicine education covers 5 years of studies with 300 ECTS.

Medical and dental medicine education is one cycle education and with its learning outcomes corresponds to/equals the Master's level. After completion of these programmes students are awarded diplomas in Medicine and Dental Medicine.

Grading System: There is a unified grading system with the highest 100 score at national level.

Admission—One of the main achievements of the higher education reform in Georgia was the establishment of a system of the Unified National Examinations. The state took a responsibility for students' admission to the first and second cycle of higher education through creating of a centralized, objective system and ensuring the principles of equity and meritocracy.

The **Quality Assurance System** in Georgia consists of internal and external quality assurance (QA) mechanisms. Internal self-evaluation is carried out by educational institutions commensurate with the procedure of evaluation of their own performance and is summarised in an annual self-evaluation report. The self-evaluation report is the basis for external quality assurance. External QA is implemented through authorization and accreditation. Authorisation is obligatory for all types of educational institutions in order to carry out educational activities and issue an educational document approved by the State. Program Accreditation is a type of external evaluation mechanism, which determines the compatibility of an educational program with the standards. State funding goes only to accredited programmes. Accreditation is mandatory for doctoral programmes and regulated professions as well as for the Georgian language and Liberal Arts. Authorisation and accreditation have to be renewed every 5 years.

The national agency implementing external QA is the Legal Entity of Public Law—National Centre for Educational Quality Enhancement (NCEQE).

4.2   Georgian Technical University

As an example of Georgian educational system a more detailed description of the Georgian Technical University (GTU) is given. The GTU is one of the biggest educational and scientific institutions in Georgia.

The main points in the history of GTU include:

1917—the Russian Emperor issued the order according to which was set up the Polytechnic Institute in Tbilisi, the first Higher Educational Institution in the Caucasian region.

1922—The Polytechnic faculty of Tbilisi State University was founded.

1928—The Departments of the polytechnic faculty merged into an independent Institute, named the Georgian Polytechnic Institute (GPI).

In the 1970s—The Institute consisted of 15 full-time and 13 part-time faculties.

1985–1987—For the volume of the advanced scientific research and work carried out by the students, the Polytechnic Institute first found a place in the USSR higher educational institutions. During this period, the Institute became the largest higher educational institution in the Caucasian region as regards the number of students (total 40,000) and academic staff (total 5000).

1990—The Georgian Polytechnic Institute was granted the university status and named the Georgian Technical University.

1995—Due to reforms and restructuring of the curriculum, GTU gradually began installation of new training standards introducing Credit System 120 of the UK credits.

2001—GTU became a full member of the European University Association—EUA.

2005—GTU joined the Bologna process and introduced 60 ECTS credits.

2005—Due to the reorganisations conducted at GTU, eight Faculties were set up.

2007—GTU was awarded accreditation by the National Center for Educational Accreditation.

2014—Due to the reorganisations conducted at GTU, ten Faculties were set up.

Educational programmes at GTU include:

- the Bachelor's Programme—240 ECTS;
- the Master's Programme—120 ECTS;
- the Doctoral Programme—180 ECTS;
- the Vocational Programme—150 ECTS.

There are four languages of study: Georgian, Russian, English and German. GTU has licensed TELL ME MORE language training software, which includes American English, British English, Dutch, French, German, Italian, and Spanish. Learning programmes are also suited to meet user needs, as the software offers Complete Beginner, Intermediate and Advanced levels.

There are ten faculties at GTU:

- the Faculty of Civil Engineering;
- the Faculty of Power Engineering and Telecommunications;
- the Faculty of Mining and Geology;
- the Faculty of Chemical Technology and Metallurgy;
- the Faculty of Architecture, Urban Planning and Design;
- the Faculty of Informatics and Control Systems;
- the Faculty of Transportation and Mechanical Engineering;
- the Business-Engineering Faculty;
- the International Design School;
- the Faculty of Agricultural Sciences and Biosystems Engineering.

Altogether 20,000 undergraduate students study in these faculties. There are also 1251 Master's students, 640 PhD students and 795 students in the Professional stream; see Table 4.2. The total amount of academic personnel at GTU is 1228, with 505 full professors, 533 associate professors, 190 assistant professors, 279 invited professors, 369 teachers and 2176 technical staff.

Table 4.2 Table of GTU higher education

Faculty	Number of undergraduate students	Number of postgraduate students	Number of doctoral students	Number of vocational students
Civil Engineering	789	90	58	221
Power Engineering and Telecommunications	1448	163	95	120
Mining and Geology	533	32	39	235
Chemical Technology and Metallurgy	850	47	66	140
Transportation and Mechanical Engineering	1493	73	85	716
Architecture, Urban Planning and Design	397	54	42	15
Business-Engineering	4195	183	48	0
Informatics and Control Systems	3413	267	152	15

4.3   STEM Programs at GTU

The GTU has been offering Engineering Degrees for decades with special attention to the following STEM fields: Computer Sciences, Computer Engineering, Energy and Electrical Engineering, Civil Engineering, Food Industry, and Forestry. The GTU participates in the Millennium Challenge Corporation (MCC) project for STEM Higher Education Development in Georgia. The project objectives are to build up capacity in Georgian public universities and to offer international standard US degrees and/or ABET (Accreditation Board for Engineering and Technology) accreditation in the STEM fields.

Three finalist consortium universities have been selected through an open competition: San Diego State University of California (SDSU); Michigan State University and University of Missouri; North Carolina State University and Auburn University.

The programme is being funded by a $29 million grant that SDSU was awarded by the MCC that entered into an agreement with the government of Georgia to improve its educational systems and infrastructure.

SDSU was one of 28 universities that competed for funding from the U.S. Millennium Challenge Corporation (MCC) to create a joint higher education programme in Georgia.

Finally, SDSU is approaching this project in partnership with Georgian Technical University, Ilia State University and Tbilisi State University—the three premier public universities in Georgia. Indicative STEM programmes include Electrical (Power) Engineering, Computer Engineering, Computer Science, Chemical Engineering and Civil Engineering fields at Georgian Technical University.

For this purpose, strengthening ABET-Accredited Georgian Degree Programmes at GTU secures better understanding of educational needs of the next generation of engineers, scientists and educators; achieving ABET accreditation of these programmes is a tangible milestone of quality improvement, providing quality of education and student outcomes of these programmes.

4.4 Mathematics at GTU

Mathematics has played and still plays nowadays a fundamental role in engineering education in GTU. In the Georgian Polytechnic Institute the Chair of Mathematics was founded in 1928. Many worldwide well-known Georgian mathematicians had been working and delivering lectures in the Georgian Polytechnic Institute, such as worldwide well-known scientists academicians Niko Muskelishvili, Ilia Vekua, Viktor Kupradze, Boris Khvedelkdze etc.

During the Soviet period all Polytechnic Institutes were forced to follow a unified mathematical curriculum with a sufficiently rich pure theoretical part. It should be mentioned that the level of the school mathematics at that time was very high in the Soviet Union and the entrants were well prepared to start learning of high mathematics, containing a very wide spectrum of courses starting from analytical geometry and classical calculus to boundary value problems for partial differential equations and theory of measure and Lebesgue integrals along with the theory of probability and mathematical statistics.

However, such a high fundamental educational level in mathematics never gave the expected and desired progress in technology and engineering. There was a big gap between theoretical preparation of students and their skills in applied practical aspects. This was one of the main drawbacks of the Soviet educational system.

In 2007, the Department of Mathematics was founded at GTU on the basis of the existing three chairs of high mathematics. The Department of Mathematics belongs to the Faculty of Informatics and Control Systems.

The staff of the Department of Mathematics consists of 20 full-time professors, 21 full-time associate professors, 3 full-time assistant professors, 7 teachers, 16 invited professors, and 5 technical specialists.

From 2008 the BSc, MSc, and PhD accredited programmes in pure and applied mathematics have been launched at the Department of Mathematics (it should be mentioned that presently the mathematical programmes are free of charge— from 2013 the Georgian Government has covered all expenses for 20 educational programmes; among them is mathematics).

The Department of Mathematics delivers lectures in high mathematics for all engineering students of GTU. Depending on the specific features of the engineering educational programmes the content of mathematical syllabuses varies and reflects mainly those parts of mathematics which are appropriate for a particular engineering specialisation.

Chapter 5
Overview of Engineering Mathematics Education for STEM in Armenia

Ishkhan Hovhannisyan

The higher education system in Armenia consists of a number of higher education institutions (HEIs), both state and private. State higher education institutions operate under the responsibility of several ministries, but most of them are under the supervision of the Ministry of Education and Science. At present there are 26 state and 41 private higher education institutions operating in the Republic of Armenia, of which 35 are accredited institutions, 6 are non-accredited institutions, 3 are branches of state HEIs and 4 are branches of private HEIs from the Commonwealth of Independent States (CIS). Higher education is provided by many types of institutions: institutes, universities, academies and a conservatory.

The University HEIs are providing higher, postgraduate and supplementary education in different branches of natural and sociological fields, science, technology, and culture, as well as providing opportunities for scientific research and studies.

The institute HEIs are conducting specialized and postgraduate academic programs and research in a number of scientific, economic and cultural branches.

The Academy (educational) HEI's activity is aimed at the development of education, science, technology and culture in an individual sphere; it conducts programs preparing and re-training highly qualified specialists in an individual field, as well as postgraduate academic programs.

The preparation of specialists for Bachelor's, Diploma Specialist's and Master's degrees, as well as postgraduate degrees, including a PhD student program (research) is implemented within the framework of Higher Education. The main education programs of higher professional education are conducted through various types of teaching: full-time, part-time or external education. The academic year, as a rule, starts on September, ending in May and is comprised of 2 semesters with 16–22 weeks of duration. There are mid-term exams and a final exam at the end of each semester.

I. Hovhannisyan (✉)
National Polytechnic University of Armenia (NPUA), Faculty of Applied Mathematics and Physics, Yerevan, Armenia

© The Author(s) 2018
S. Pohjolainen et al. (eds.), *Modern Mathematics Education for Engineering Curricula in Europe*, https://doi.org/10.1007/978-3-319-71416-5_5

ARMENIAN EDUCATION SYSTEM DIAGRAM

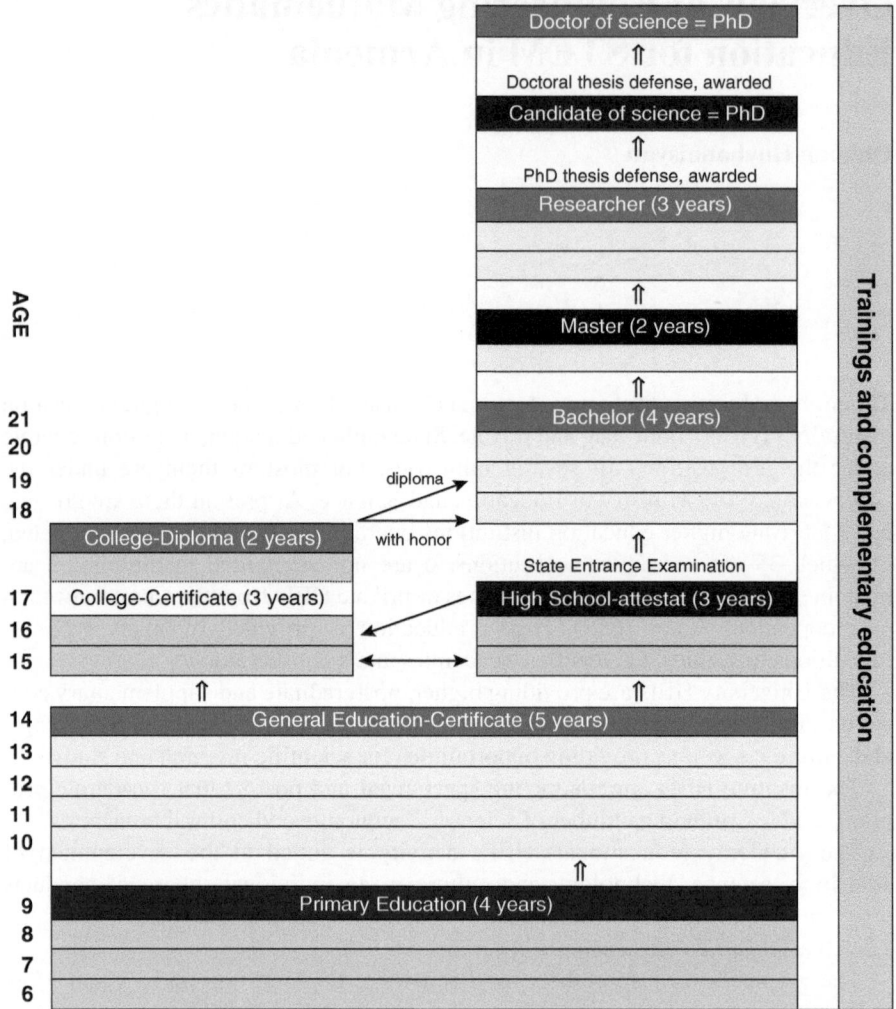

Fig. 5.1 The Armenian education system

The formal weekly workload (contact hours) that students are expected to carry out depends on the type of programs and differs considerably from institution to institution within the country, but common practices are as follows: for Bachelor programs 28–32 h per week (sometimes up to 36), for Master programs 16–18 h and for postgraduate (Doctorate) programs 4–8.

Starting from 2008 all educational programs in Armenia are based on the ECTS. Internal systems of student evaluation and assessment are regulated by the HEIs themselves. Students' learning outcomes are assessed on the basis of examinations and tests, which are conducted in writing or orally. The results of examinations are assessed by grading systems varying considerably among institutions (5-, 10-, 20- or 100-point marking scales, 4 scale A–F letter grading, etc.). A final evaluation of graduates in state HEIs is conducted by state examination committees both through the comprehensive examination on specialization and defense of graduation work (diploma project, thesis or dissertation) or schemes only one of them is used.

A diagram of the Armenian education system can be seen in Fig. 5.1.

5.1   National Polytechnic University of Armenia

The National Polytechnic University of Armenia (NPUA) is the largest engineering institution in Armenia.

The University has a leading role in reforming the higher education system in Armenia. NPUA was the first HEI in RA that introduced three level higher education systems, implemented the European Credit Transfer System (ECTS) in accordance with the developments of the Bologna Process.

The University is a member of European University Association (EUA), the Mediterranean Universities Network, and the Black Sea Universities Network. It is also involved in many European and other international academic and research programs. The University aspires to become an institution, where the entrance and educational resources are accessible to diverse social and age groups of learners, to both local and international students, as well as to become an institution which is guided by a global prospective and moves toward internationalization and European integration of its educational and research systems.

5.2   Overview of Mathematics Education at NPUA

The NPUA Faculty of Applied Mathematics and Physics is responsible for major and minor mathematical education at the University. It was established in 1992 by uniting University's 3 chairs of Higher Mathematics. Academician of National Academy of Sciences Prof. Vanik Zakaryan is the founder-Dean of the Faculty. Nowadays the faculty is one of the top centers of Mathematics and Physics in the country and the biggest faculty in the University, having more than 90 full-time faculty members (12 professors and 48 associate professors). The student body of the faculty consists of approximately 200 majors (all programs) and more than 3000 minors. The faculty offers the following programs as majors:

- Bachelor in Informatics and Applied Mathematics;
- Bachelor in Applied Mathematics and Physics;
- Master in Informatics and Applied Mathematics;
- PhD in Mathematics.

In addition to these major programs, the Faculty caters to the mathematics and physics subsidiary (minor) courses in other BSc and MSc programs of the University with specializations in Engineering, Industrial Economics and Management. It also renders services of its full-time faculty to teach elective courses of mathematics at MSc and PhD programs.

5.3 Mathematics Courses at NPUA

The Bachelors of Science in Engineering (BSE) degree at NPUA involves completing 240 credit hours of courses in various categories, which include: a module of languages, Economics and Humanities module, a module of General Engineering courses, Module of Specialization courses, and a module of Math and Natural Science courses. The University requires that all engineering students, regardless of their proposed engineering major, complete specific courses in the core subjects of mathematics which are listed in Table 5.1 with the number of ECTS credits for each course.

Table 5.1 BSc mathematics courses at NPUA

Course	Semester	Credits	Hours
Mathematical Analysis 1 (An introduction to the concepts of limit, continuity and derivative, mean value theorem, and applications of derivatives such as velocity, acceleration, maximization, and curve sketching)	1	5	Lectures 32, tutorials 32
Mathematical Analysis 2 (introduction to the Riemann integral, methods of integration, applications of the integral, functions of several variables, partial derivatives, line, surface and volume integrals)	2	5	Lectures 32, tutorials 32
Analytic Geometry and Linear Algebra (vector spaces and matrix algebra, matrices and determinants, systems of linear equations)	1	4	Lectures 32, tutorials 16
Theory of Probability and Statistical Methods (probability space axioms; random variables and their distributions, expectation values and other characteristics of distributions)	3	4	Lectures 32, tutorials 16
Discrete Mathematics	2	2	Lectures 32, tutorials 16

Table 5.2 MSc mathematics courses at NPUA

Course	Semester	Credits	Hours
Discrete Mathematics	1	5	Lectures 48, practices 16
Numerical Methods	1	5	Lectures 48, practices 16
Mathematical Programming	1	5	Lectures 48, practices 16
Linear Algebra	1	5	Lectures 48, practices 16
Theory of Probability and Statistical Methods	1	5	Lectures 48, practices 16
Functions Approximation by Polynomials	1	5	Lectures 48, practices 16

In addition to the core mathematics courses, the Master of Science in Engineering (MSE) degree requires students to complete at least 5 credits of advanced mathematics elective courses. Table 5.2 contains the list of advanced mathematics elective courses for MSE students.

Chapter 6
Overview of Engineering Mathematics Education for STEM in EU

6.1 Engineering Mathematics Education in Finland

Tuomas Myllykoski and Seppo Pohjolainen (✉)
Tampere University of Technology (TUT), Laboratory of Mathematics, Tampere, Finland
e-mail: tuomas.myllykoski@tut.fi; seppo.pohjolainen@tut.fi

The Finnish higher education system consists of two complementary sectors: polytechnics and universities. The mission of universities is to conduct scientific research and provide undergraduate and postgraduate education based on it. Polytechnics train professionals in response to labor market needs and conduct R&D which supports instruction and promotes regional development in particular.

Universities promote free research and scientific and artistic education, provide higher education based on research, and educate students to serve their country and humanity. In carrying out this mission, universities must interact with the surrounding society and strengthen the impact of research findings and artistic activities on society.

Under the new Universities Act, which was passed by Parliament in June 2009, Finnish universities are independent corporations under public law or foundations under private law (Foundations Act). The universities have operated in their new form from 1 January 2010 onwards. Their operation is built on the freedom of education and research and university autonomy.

Universities confer Bachelor's and Master's degrees, and postgraduate Licentiate and Doctoral degrees. Universities work in cooperation with the surrounding society and promote the social impact of research findings. The higher education

© The Author(s) 2018
S. Pohjolainen et al. (eds.), *Modern Mathematics Education for Engineering Curricula in Europe*, https://doi.org/10.1007/978-3-319-71416-5_6

system, which comprises universities and polytechnics, is being developed as an internationally competitive entity capable of responding flexibly to national and regional needs. There are technological universities in Helsinki, Tampere and Lappeenranta, and technological faculties in Oulu, Vaasa and Turku. Approximately 4000–5000 students begin their studies in one of these universities annually. The Finnish education system diagram is in Fig. 6.1.

The system of polytechnics is still fairly new. The first polytechnics started to operate on a trial basis in the beginning of 1990s and the first was made permanent in 1996. By 2000 all polytechnics were working on a permanent basis. Polytechnics are multi-field regional institutions focusing on contacts with working life and on regional development. The total number of young and mature polytechnic students is 130,000. Polytechnics award over 20,000 polytechnic degrees and 200 polytechnic Master's degrees annually. The system of higher degrees was put in place after a trial period in 2005 and the number of polytechnic Master's programs is expected to grow in the coming years.

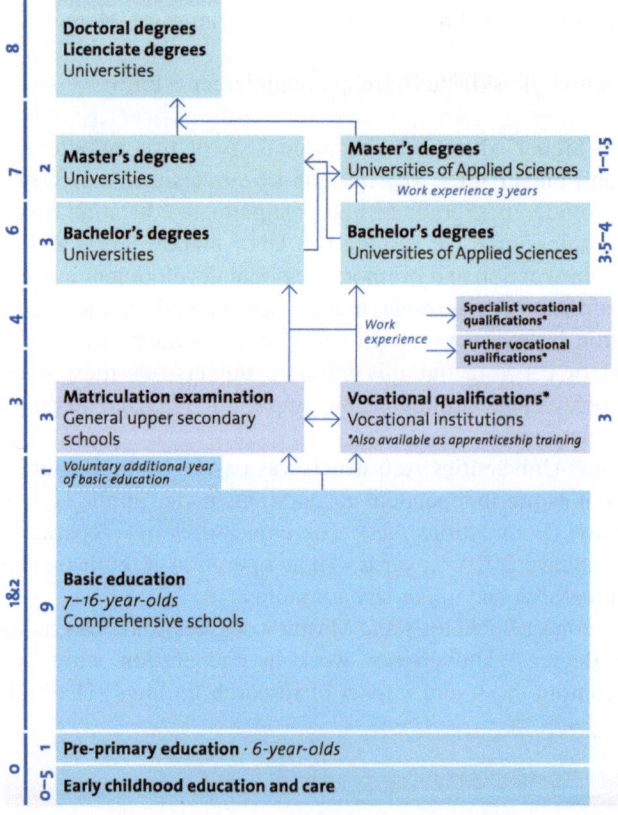

Fig. 6.1 The Finnish education system (http://www.oph.fi/english/education_system)

6.1.1 Tampere University of Technology

Tampere University of Technology (TUT) is Finland's second-largest university in engineering sciences. TUT conducts research in the fields of technology and architecture and provides higher education based on this research. TUT is located in Tampere, the Nordic countries' largest inland city, some 170 km north of the capital Helsinki. TUT's campus in the suburb of Hervanta is a community of 10,500 undergraduate and postgraduate students and close to 2000 employees. Internationality it is an inherent part of all the University's activities. Around 1500 foreign nationals from more than 60 countries work or pursue studies at TUT.

TUT offers its students an opportunity for a broad, cross-disciplinary education. Competent Masters of Science of Technology and Architecture as well as Doctors of Technology and Philosophy graduated from TUT are in high demand among employers.

The University combines a strong tradition of research in the fields of natural sciences and engineering with research related to industry and business. Technology is the key to addressing global challenges. The University's leading-edge fields of research are signal processing, optics and photonics, intelligent machines, biomodeling and the built environment.

TUT generates research knowledge and competence for the benefit of society. The University is a sought-after partner for collaborative research and development projects with business and industry and a fertile breeding ground for innovation and new research- and knowledge-based companies.

In 2013 the total funding of TUT Foundation, which operates as Tampere University of Technology, was 157.6 million euros. Close to 50% of the University's funding was external funding, such as revenue from The Finnish Funding Agency for Technology and Innovation (Tekes), industry, the Academy of Finland and EU projects.

TUT started operating in the form of a foundation in the beginning of 2010. The independence of a foundation university and the proceeds of the 137 million euro foundation capital further promote the development of research and education at TUT.

6.1.2 Overview of Mathematics Education at TUT

The Department of Mathematics is responsible for teaching core mathematics to all engineering students at Tampere University of Technology, and it offers courses and degree programs at the Bachelor's, Master's, and Postgraduate level for studies in mathematics. The faculty of the department conducts research in mathematics and its applications at an internationally competitive level.

The Department of Mathematics offers mathematics and statistics expertise for research and development projects in the private and public sectors. Research services and collaboration can range from informal working relationships to shorter- or longer-term research contracts.

The teaching of mathematics for the following degree programs is the responsibility of the Department of Mathematics on the Bachelor and Master level:

- Automation Engineering
- Biotechnology
- Civil Engineering
- Signal Processing and Communications Engineering
- Electrical Engineering
- Environmental and Energy Technology
- Industrial Engineering and Management
- Information and Knowledge Management
- Information Technology
- Materials Engineering
- Mechanical Engineering
- Science and Engineering

Generally, the mandatory amount of mathematics is 27 ECTS in the Bachelor's degree. The mandatory mathematics consists of Engineering Mathematics 1–3 courses (altogether 15 ECTS) (Table 6.1), studied during the first year, and three elective courses 4 ECTS each on the first and second year (Table 6.2). The degree program makes the recommendation on the elective courses suitable for their students. The degree program can place one or two mandatory mathematics courses in their Master's program. Then these courses are not part of the BSc program but belong to the MSc program. The degree programs may recommend students to include additional mathematics courses in their study plan. In this case the amount of mathematics exceeds 27 ECTS.

Table 6.1 Mandatory engineering mathematics courses at TUT

Course	ECTS	Year	Contents
Engineering Mathematics 1 (EM1)	5	1	Set theory and mathematical logic; real functions; elementary functions; limits; derivative; complex numbers; zeros of polynomials
Engineering Mathematics 2 (EM2)	5	1	Vectors in $\mathbb{R}^n$ spaces; linear equations, Gauss' elimination; vector spaces; matrices, eigenvectors, determinants; orthogonality
Engineering Mathematics 3 (EM3)	5	1	Indefinite and definite integral; first and second order differential equations; sequences; series

Table 6.2 Elective engineering mathematics courses at TUT

Course	ECTS	Year	Contents
Engineering Mathematics 4 (EM1)	4	1	Multivariable functions, limit, continuity, partial derivatives, gradient; vector valued functions, matrix derivative, chain rule; maxima, minima, Lagrange's method; plane and space integrals
Algorithm Mathematics (AM)	4	2	Set theory, methods of proofs; relations and functions; propositional and predicative logic; induction and recursion; Boolean algebra
Discrete Mathematics (DM)	4	1, 2	Step-, impulse-, floor-, ceiling- signum functions; Z-transform, difference equations; number theory; graph theory
Fourier Methods (FM)	4	2	Real Fourier series; linearity, derivation; complex Fourier series; discrete Fourier transform
Operational Research (OA)	4	2	Linear optimization; Simplex-method, sensitivity; dual problem; transport model with applications; warehouse models; game theory
Probability Calculus (PC)	4	2	Random variable and probability, Bayes' formula; distributions and their parameters; joint distributions, central limit theorem
Statistics (MS)	4	2	Descriptive statistics, samples; hypothesis testing, parametric and nonparametric cases
Vector Analysis (VA)	4	2	Gradient, divergence; line integrals; conservative vector field; surfaces, area, surface integral, flux and Gauss' theorem

The study program recommends the selection of (at least) three of these courses

6.1.2.1   Major/Minor in Mathematics (BSc, MSc)

In addition to the courses provided for all degree programs, the department also has a well-rounded group of students who study mathematics as their major. The courses provided for these students are often based on the research topics of the department, as this further develops the department's strategy.

The Department of Mathematics offers courses and degree programs at the Bachelor's, Master's, and at the Postgraduate level. Doctoral studies can be carried out in the main research areas of the department.

Currently there are two majors in the Master's program of Science and Engineering fully given in English: Mathematics with Applications and Theoretical Computer Science, both by the Department of Mathematics. These offer the uniquely useful combination of strong mathematical modeling and tools of logical and algorithmic analysis. There is also a minor subject in Mathematics. As per agreement, it may also be included in other international Master's programs at TUT.

The optional focus areas in Mathematics with Applications are Analysis, Discrete Mathematics, and Mathematical and Semantic Modeling. The subject can also be chosen as an extended one.

The postgraduate studies program leads to the PhD degree or the degree of Doctor of Technology. Subjects of the postgraduate studies at the department usually follow the research topics of the research groups. There is a local graduate school, which can provide financial support for doctoral studies.

On the Bachelor level, the minor or major in mathematics consists of 25 ECTS in mathematics in addition to the 27 ECTS studied by all students. For the major, the students will also complete a Bachelor's thesis worth 8 ECTS. If the student chooses not to take mathematics as a major, then they must complete an additional 10 ECTS of mathematics for a total of 60 ECTS. In the Master phase of studies, the students will complete either 30 or 50 ECTS of mathematics for their major, depending on their choice of minor studies. Those meaning to graduate from the Master's program with a major in mathematics are required to write the Master's thesis in mathematics that is worth 30 ECTS.

6.1.2.2   Teacher Studies at TUT in Mathematics

The students are offered the possibility of studying mathematics with the goal of attaining competency for teaching mathematics at Finnish schools. Students must study a minimum of 50 ECTS of mathematics in the university to be able to apply for the program. After applying, the students are evaluated by a board of academics at University of Tampere. Those who pass evaluation are given the possibility of studying at University of Tampere to obtain 60 ECTS of mandatory pedagogical studies. Students will study both at TUT and at the Tampereen Normaalikoulu—high school—where they work as real teachers with real students. The major in mathematics for teacher students is 120 ECTS, a minor is 60 ECTS of mathematics. Teacher students will also complete a 60 ECTS minor in pedagogical studies. Students have a possibility of studying multiple sciences in their teacher studies, with mathematics being accompanied by chemistry, physics and information technology. It is often suggested that students choose multiple sciences in order to further develop their possibilities in the future when looking for a job. The overall structure of teacher studies is depicted in Fig. 6.2.

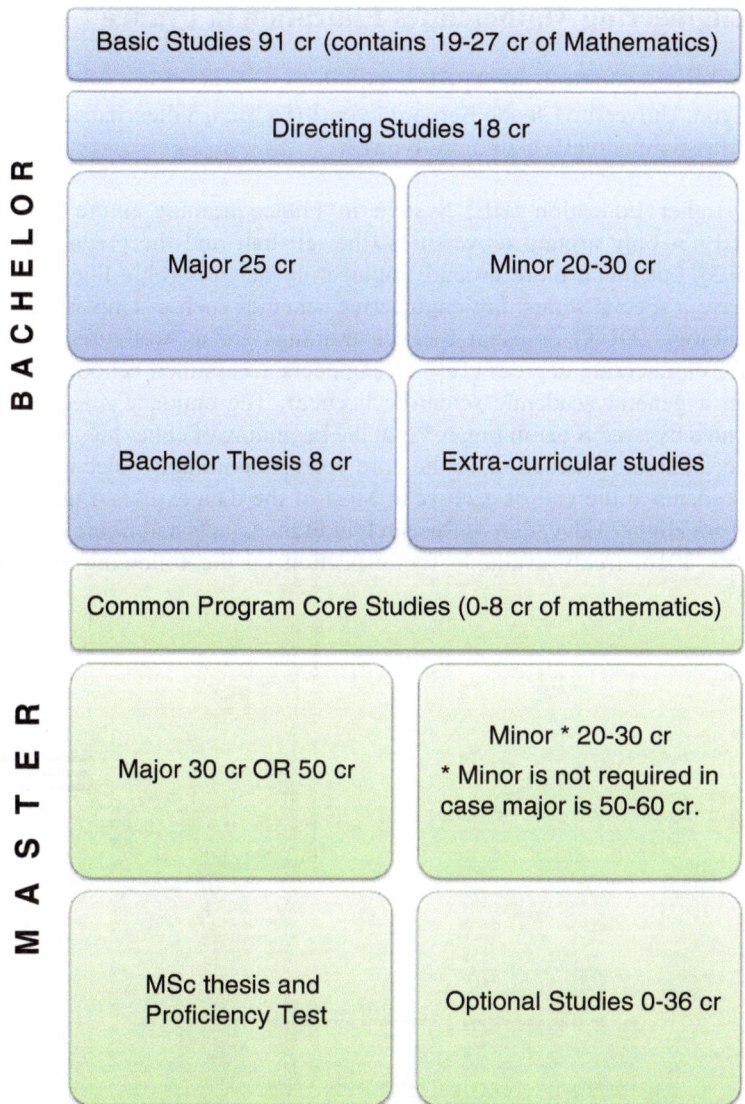

Fig. 6.2 Teacher studies in mathematics at TUT

6.2   Engineering Mathematics Education in France

Christian Mercat and Mohamed El-Demerdash
IREM Lyon, Université Claude Bernard Lyon 1 (UCBL), Villeurbanne, France
e-mail: christian.mercat@math.univ-lyon1.fr

The Higher Education (HE) System in France training future engineers is organized not only around universities (the left half and the green vertical line in Fig. 6.3), but much more around Engineering Schools (blue line to the right) which have a special status. Pre-engineering schemes such as University Diploma of Technology (DUT) or usual Licence trainings are as well used as stepping stones for engineering degrees (16% of engineers access their school with a DUT, 6% from a general academic scientific licence). The entrance selection scheme (represented by a red S bar in Fig. 6.3 is at the beginning of either the first year, right after Baccalaureate graduation, or the third year. A continuous selection weeds out failing students at the end of each year. Most of the data exploited in this chapter comes from Higher Education & Research in France, facts and figures, 9th edition, November 2016, freely available for inspection on the following governmental website[1]:

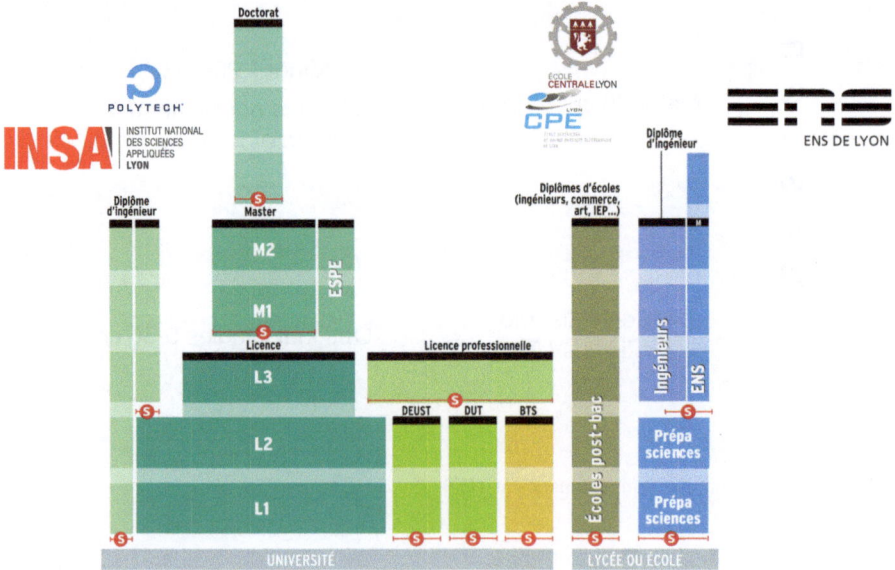

Fig. 6.3 French education system

6.2.1   Universities

There are 73 universities in France, for a total of 1.5 million students, which represent 60% of the number of Higher Education students; see Fig. 6.4. The number of students has increased by a factor of 8 in the last 50 years to reach 2.5 million, the proportion of Baccalaureate holders increasing from a third in 1987 to two-thirds of a generation in 1995 and three-quarters nowadays. The demographical increase is expected to make the numbers of HE students steadily grow in 10 years to reach 2.8 M. Short technician diplomas, BTS and DUT, are mainly responsible for this increase. These short technical diplomas follow the creation of vocational and technological Baccalaureates; see Fig. 6.5.

Whereas any European freshmen can enter French university (there is no entrance selection), a significant number of HE students will never go to university and it

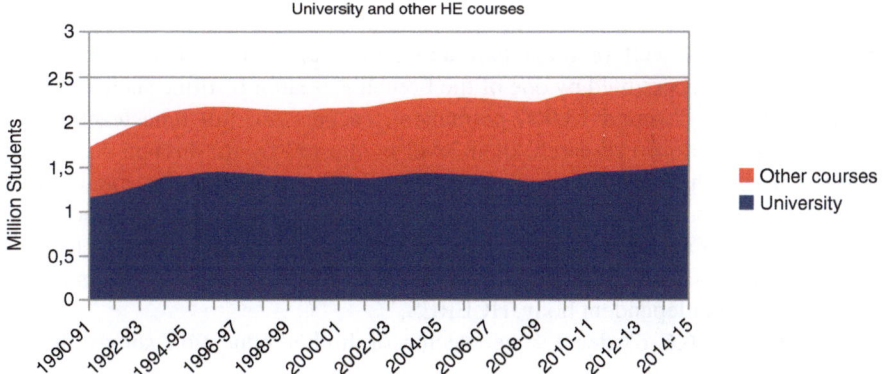

Fig. 6.4   The number of students in Higher Education

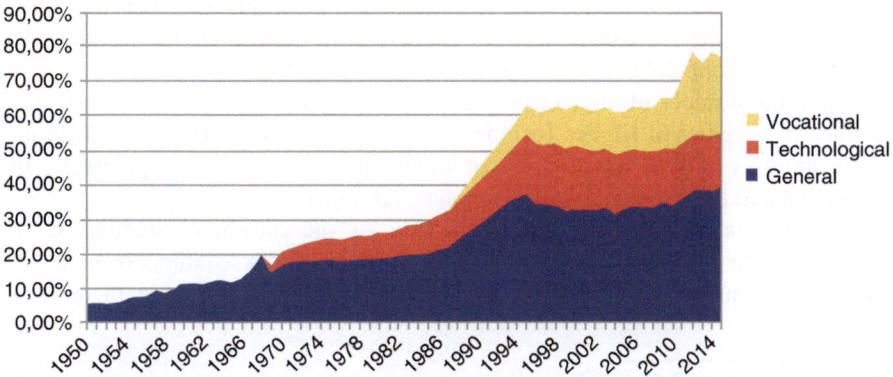

Fig. 6.5   Percent of a generation with Baccalaureate

Fig. 6.6 Degrees per sector
of Higher Education

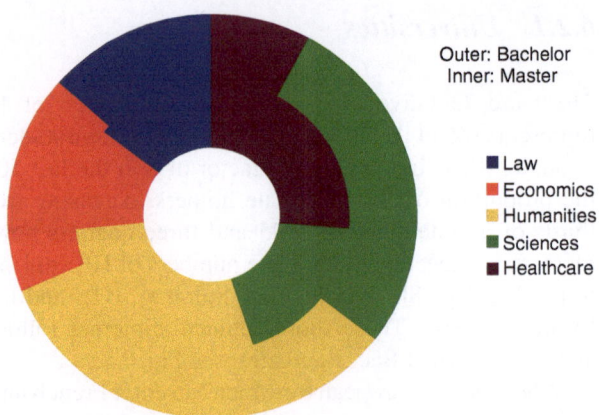

Outer: Bachelor
Inner: Master

■ Law
■ Economics
■ Humanities
■ Sciences
■ Healthcare

is especially true for engineers. There are around 63,000 permanent HE teachers in France, half of them in science and technology departments. Around 50,000 are researchers as well (a fixed half and half loads) and usually belong to a research institute accredited by one of the French Research Institute such as CNRS, INRIA or CEA.[2] Around 13,000 permanent teachers are full time teachers and are not supposed to do research, their level is attested by the French Agrégation or equivalent. These Professeurs Agrégés (PRAG) are especially numerous in engineers schools. Around 24,000 teachers have non-permanent teaching positions. The curriculum is accredited by the ministry of higher education and research. These research institutes, as well as the programs and degrees delivered by universities are evaluated by an independent body, HCERES.[3]

The plot Fig. 6.6 of degrees per sector of higher education shows that, in proportion with other sectors, many more students leave Sciences with a simple Bachelor's degree and will not achieve a Master's degree. The reasons are twofold, a good and a bad one: first, a Bachelor's degree in science is sufficient to get a job compared to humanities or law for example, especially the BTS and DUT, and, second, science students are more likely to drop out earlier. Two-thirds of the scientific and technical Masters are in fact engineering degrees.

While the number of engineer's degrees has slightly increased, the lower pre-engineering degrees of Higher Technician Diploma (BTS) and University Diploma of Technology (DUT) have increased much more, following, respectively, the vocational and technological Baccalaureates; see Fig. 6.7.

Foreign students account for about 15% of the university students (DUT included) with a sharp increase in the first years of this century, from 8% in 2000. Some courses, such as preparatory schools (CPGE) and Technical University

---

[2]National Center for Scientific Research http://www.cnrs.fr/; National Institute for computer science and applied mathematics http://www.inria.fr; French Alternative Energies and Atomic Energy Commission http://www.cea.fr.

[3] High Council for Evaluation and Research and Higher Education, http://www.hceres.fr/.

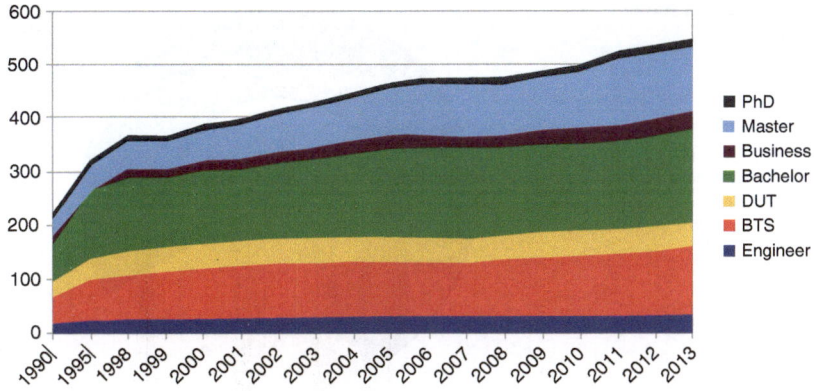

Fig. 6.7 Numbers of degrees (in thousands)

Diplomas (DUT), having no foreign counterparts, do not attract foreign students although they are competitive and channel a lot of engineers in France, especially the top engineering schools which have de facto mainly French students. Parallel admission systems allow for the inclusion of foreign students. It is especially the case for foreign students already holding a Master's degree and training a specialization in a French engineering school; see Fig. 6.8.

These students come traditionally mainly from Africa, notably Morocco, Algeria, Tunisia and Senegal, but more and more from Asia, remarkably from China. Germany and Italy are the main European partners; see Fig. 6.9.

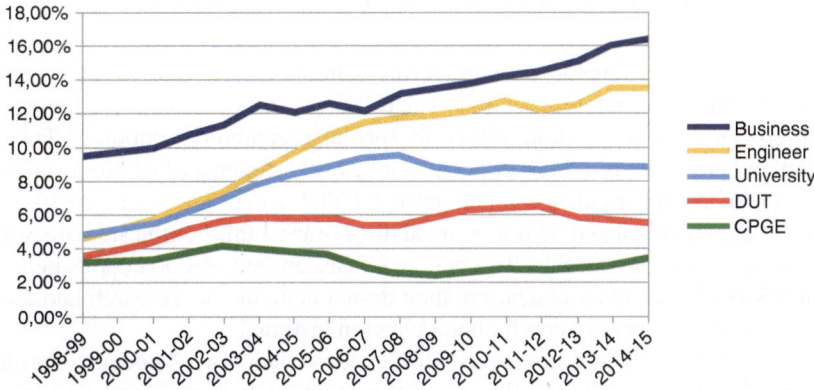

Fig. 6.8 Percentage of foreign students

Fig. 6.9 Origin of foreign students

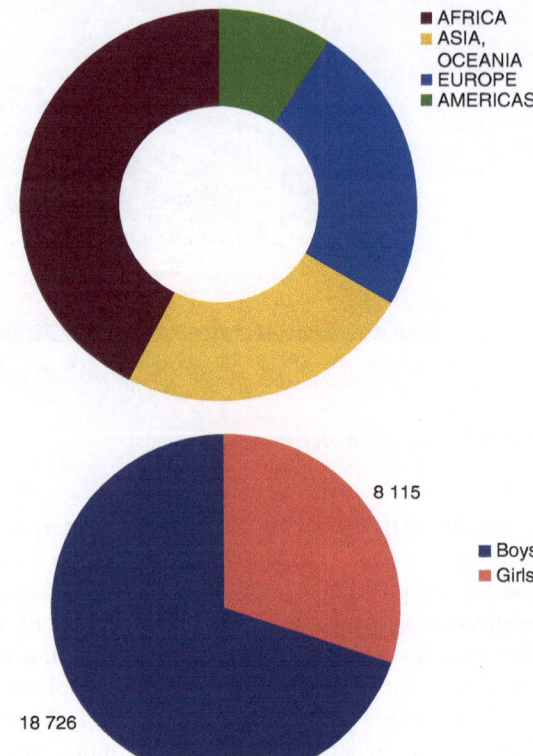

Fig. 6.10 Population and gender repartition in Scientific Preparatory Schools (CPGE) in 2016

6.2.2 *Preparatory Courses (CPGE and Internal)*

The main entrance scheme to engineering schools (more than 80%) is not via university but via preparatory schools. The two first years of the engineering training are preparatory courses, done either in special preparatory schools (CPGE) or already integrated within the school. Around 27,000 students (93% from scientific, 7% from technical Baccalaureate) go through CPGE each year; see Fig. 6.10. There are great geographical and demographical disparities: a third of the students studies in Paris and a third are female. Teachers in CPGE, around 86,000 people, belong to the ministry of secondary education, they do not conduct any research and are not affiliated with a university or a higher education institute.

The main networks of engineers schools have their own "students shuffling" system, allowing students to allocate, at the completion of their two first internal preparatory years, according to their accomplishments and desires, especially the Polytech and INSA networks.

At the end of these 2 years, competitive exams, national or local, rank students, in different ranking systems. Different schools unite in consortia and networks, sharing the same examinations, allowing themselves to rank internally students in order for them to express their choices and allocate places according to offer and demand. Some of these examinations are incompatible with one another, so students have to choose which schools they want to apply to. The three main public competitive examinations are: Polytechnique-Écoles Normales-ESPCI, Centrale-Supélec, Mines-Ponts. These exams are open to students having followed 2 years of preparatory schools, whether at university (a small proportion) or in special training schools (CPGE).

6.2.3 Technical University Institutes (IUT) and Their Diploma (DUT)

With a Technical Baccalaureate, students can enter university and become higher technicians in Technical University Institutes (113 IUT in France and one in UCBL), which are limited to 2 years training, eventually followed by a professional license. More than 70% of them receive a University Diploma in Technology (DUT), an intermediate degree of the LMD system. This diploma allows for immediate employability and about 90% of the DUT students after graduation do get a permanent position within the first year. But only 10% of the graduates choose to do so! DUT is somehow "hijacked" by 90% of students that in fact are looking for further training, and especially as engineers. The French job market lacks skilled technicians (1500 euro/month median first salary): students want to invest in studies in order to get a Master (1900 euro/month median first salary) or an engineer's salary (2700 euro/month median first salary) for the same 90% students placement success rate. This situation has to be kept in mind when analyzing the French Engineering School system.

6.2.4 Engineering Schools

Around 100,000 engineers are being trained in 210 French engineering schools today, leading to 33 thousand graduations a year, a fifth of them in schools included in a university. Unlike the general Bachelor program at university, the entrance to these schools is competitive and the dropout rate is very low.

An engineering school has to be accredited by the Ministry of Higher Education and Research, after an inquiry, every 6 years by a special quality assessment body, the Engineering Accreditation Institution (CTI); this requirement has existed since 1934. Students training, students job placement, recruitment of the personnel, industrial and academic partnerships, and self quality assessment are among the main criteria. CTI belongs to the European Association for Quality Assurance in Higher Education (ENQA) and the European Consortium for Accreditation (ECA).

These schools are often independent and usually do not belong to a university. In particular, engineering school training is limited up to the fifth year after national

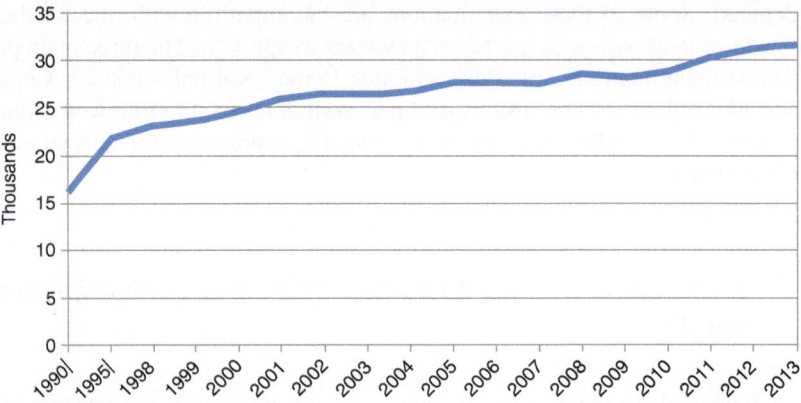

Fig. 6.11 Number of engineering degrees (in thousands)

Baccalaureate Degree or equivalent, that is to say, the equivalent of Master degrees, and these institutions are not permitted to deliver Doctoral diplomas. In order to do so, they have to establish a partnership with a university. A fair amount of their staff usually belongs to a research laboratory hosted by a university. Therefore, compared to partner countries such as Finland or Russia, most universities are "classical" universities, while engineering schools take the role of "technical" universities. Moreover, let us recall that in most of these schools, specialized work begins in the third year, the first two being preparatory years, taught internally or in another institution (CPGE). All in all, engineering degrees represent around two-thirds of all the scientific and technical Masters in France.

The 210 accredited engineering schools in France fall into three main categories. First, there are schools integrated inside a university, Polytech is an example of such schools. Then there are schools with integrated preparatory years, INSA is an example. Then there are schools which train only for 3 years after 2 years of preparatory school (CPGE), École Centrale is an example. We will further describe these paradigmatic examples below. Around a third of the engineering schools are private, for a fourth of engineering students. As shown in Fig. 6.11 the number of engineering degrees awarded each year has been increasing by 3.4% a year over the last 5 years, reaching 32,800 in 2015.

Around 4% of engineering students continue their studies with a PhD. There are around 1% of the students doing so in private schools and professional Masters, compared with 14% in research Masters.

The engineering schools are often part of networks such as the Polytech[4] or the INSA[5] networks. The Polytech network is constituted of 13 public schools, training more than 68,000 students and leading to 3000 graduations per year. The INSA

---

[4]http://www.polytech-reseau.org/.

[5]http://www.groupe-insa.fr/.

network of six schools trains 2300 engineers per year for 60 years, amounting to more than 80,000 engineers on duty today.

Around 100,000 students are being trained in 210 French engineering schools today, for 33 thousand graduations a year, a fifth of them in schools included in a university. Unlike universities, the entrance to these schools is competitive. Their curricula and diplomas are continuously assessed and validated by the Commission of Engineer Title (CTI). This commission is a member of the European Association for Quality Assurance in Higher Education and belongs to the French Ministry of Higher Education and Research.

The entrance selection scheme is either at the beginning of the first year, right after Baccalaureate graduation, or in the third year. A continuous selection weeds out failing students at the end of each year. The two first years are preparatory courses, done either in special preparatory schools (CPGE) or already integrated within the school.

6.2.5   Engineer Training in Lyon

Higher Education players in Lyon are gathered into the Université de Lyon consortium, which comprises 129,000 students and 11,500 researchers; see Fig. 6.12.

Université Claude Bernard Lyon is the science and technology university of the Université de Lyon. There are 3000 researchers for 68 research laboratories and 40,000 students in 13 teaching departments. The mathematics research laboratory is the Institut Camille Jordan (ICJ UMR 5208 CNRS, 200 members). Most of the mathematics teachers from the neighboring Engineers Schools, INSA, École Centrale, Polytech, which are as well researchers, belong to this research institute. Lyon has another smaller research institute in the École Normale Supérieure (UMPA UMR 5669 CNRS, 50 members) hosting much less applied mathematicians and no engineers trainer. CPE researchers belong to the Hubert Curien laboratory.

6.2.5.1   Civil Engineering

Course Description: The course Civil Engineering and Construction is the third year of Mechanics License—Civil Engineering. It is administratively attached to the Department of Mechanics Faculty of Lyon 1.

Training Duration: 2 semesters. Number of hours of training at the University: 600 h. Period of internship: 6 weeks. According to the student profile, specific modules and differentiated lessons are implemented.

Course Overview: The aim of this license is to provide an operational, flexible and scalable framework combining scientific and technological knowledge in the area of Building and Public Works. All major areas of construction are discussed: Drawing, Work Management, Energy, Structures, Soil Mechanics, Topography, Materials, etc.

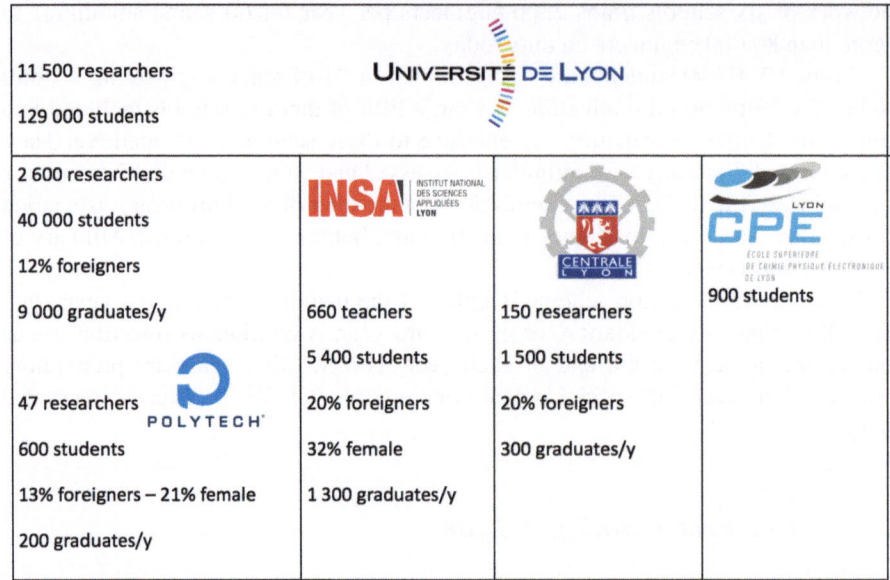

Fig. 6.12 Université de Lyon Consortium

6.2.5.2 Master Program

Courses offered at the Master's degree level satisfy a dual objective of preparing students for research and providing courses leading to high level professional integration. The Master's degree is awarded after acquisition of 120 credits after the "license" (Bachelor's degree) on the basis of training organized in four semesters. The first 60 credits (M1) can, by request of the student, receive an intermediate level national "maîtrise" diploma, a heritage of the previous French HE system. The remaining credits lead to the awarding of the national "Master" diploma.

6.2.5.3 Polytech

UCBL hosts the engineers school Polytech. It belongs to a network of 13 engineering schools embedded into universities. It represents a pretty new trend in France, the school in Lyon was founded in 1992 and joined the Polytech network in 2009. It grew to become quite an alternative to more classical engineering schools. The recruitment of Baccalaureate students (aged around 18) is done at a national level through a common procedure, shared with other 29 engineering schools: the Geipi Polytech competitive exam awarding more than 3000 students a ranking into the affiliated schools from which to make a choice. Polytech Lyon majors are chosen by around 200 of them per year (Fig. 6.13).

Over 14 000 students
85 Majors grouped in 5 big scientifc areas

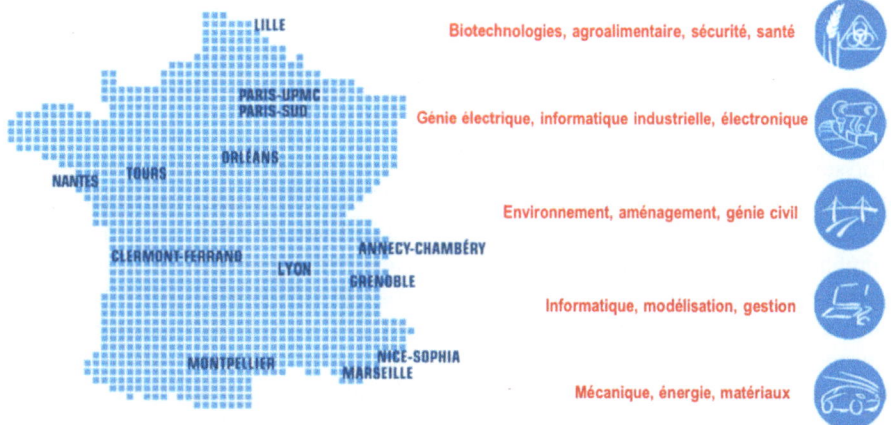

Biotechnologies, agroalimentaire, sécurité, santé

Génie électrique, informatique industrielle, électronique

Environnement, aménagement, génie civil

Informatique, modélisation, gestion

Mécanique, énergie, matériaux

Fig. 6.13 The network of Polytech universities

The two first years are preparatory, following the program of other classical preparatory schools but at the university. The actual choice of the engineer school is done at the completion of these 2 years. A reshuffling of the students, according to their choice of major topic and achievements, is performed inside the Polytech network. A student having begun studying in Lyon might very well end up finishing her/his studies in Polytech Grenoble, for example, because she/he grew interested in the Geotechnic major prepared there. But majors in Lyon are attractive so the reverse is the case: from 120 preparatory students, numbers jump to around 200 in the third year (L3) when majors begin. These majors are selective and it fuels the competition between students at a national level in the network. The sex ratio is around 20% of girls, but a great effort is being made in that respect, the preparatory school ratio being now about 30%.

There are six majors in Polytech Lyon, rooted in the scientific workforce in UCBL:

- Biomedical engineering;
- Computer science;
- Materials sciences;
- Modeling and applied mathematics;
- Mechanical engineering;
- Industrial engineering and robotics.

Every student from the third year on belongs to one of them. These majors are backed up by research laboratories of the UCBL to which the teachers belong as researchers. Most of these laboratories are associated with CNRS (French National

Agency for Research). For mathematics, it is the Institut Camille Jordan UMR CNRS 5208 (ICJ). Two 6-month internships in industry or research laboratory are performed in the fourth and fifth year of study. International training is mandatory, whether as a student or as an intern.

6.2.5.4   INSA

The seven National Institutes for Applied Sciences (INSA) form the largest network of engineer schools in France, amounting to 12% (10,200 students) of all the trained French engineers. The first institute was open in Lyon in 1957 and had a special emphasis on social opportunities, humanities and international cooperation: special cycles prepare engineer students in the spirit of cultural openness and bilingualism such as EURINSA, ASINSA, AMERINSA, NORGINSA, SCAN, respectively, for European, Asian, South-American, Nordic and English speaking students. French students and foreign students are mixed half and half. Each learns the language of the other group during the two first preparatory years, followed by a 1 month industrial internship in the alien culture. The cultural and ethical dimensions of science and technology, and their teaching, are therefore a distinctive point of these institutes. Just as in the case of Polytech, the two first years are preparatory and students can or may have to move from one INSA to another, at the end of the second year, in order to find a major adapted to their achievements and desires. The INSA in Lyon counts among its 5400 students over 5 years of training, 20% of foreign students, 32% of female students and 31% of grant students, which is a distinctive mark of its social openness, compared to other more socially discriminative schools. Among its 660 teachers, more than half are researchers as well and depend on a laboratory. In mathematics, among the 45 members of the math Pole, 13 belong to ICJ, and another 15 to other laboratories (computer science, acoustics, energy, civil engineering).

There are 12 majors taught in Lyon, linked to research laboratories. They begin in the third year:

- Biochemistry and Biotechnologies
- Bio-computer Sciences and Modeling
- Civil Engineering and Urbanism
- Electrical Engineering
- Energetic and Environmental Engineering
- Mechanic Conception Engineering
- Mechanic Development Engineering
- Plastic Process Mechanic Engineering
- Industrial Engineering
- Computer Science
- Material Sciences Engineering
- Telecommunications

6.2.5.5 École Centrale de Lyon

This engineer school was founded in 1857 and belongs to the "Centrale group" of eight schools, five of which are in France, others in Beijing (China), Casablanca (Morocco) and Mahindra (India). This network selects students after 2 years of preparatory school. Teaching therefore only lasts 3 years. The national competitive exam, common to 10 schools (the five French Centrale and five other schools) is called Concours Centrale-Supélec, is one of the three top competitive exams in France with Polytechnique-ÉNS-ESPCI and Mines-Ponts. The best students in French preparatory schools take these three selection exams. There are around 1000 students in this 3-year school, completed with around 100 Master students and 200 PhD and post-doctoral students. Half of the 200 teachers are as well researchers. The 3 years of study are divided into 2 years of common core followed by seven majors lasting for one semester:

- Civil Engineering and Environment;
- Mathematics and Decision;
- Aeronautics;
- Transportation and Traffic;
- Computer Science and Communication;
- Energy;
- Micro- and Nano-biotechnologies.

The study is structured around three main collaborative projects, one each year, of 9 months in the first 2 years and of 6 months in the third year. Internships punctuate the education as well, of increasing complexity, from an execution internship of 1 month in the first year, application of 3–4 months in the second year, to the study internship of 6 months in the third year. Sabbatical leave during the school is promoted, for personal projects, such as industrial internship or academic study abroad or professionally oriented projects. Once graduated, students can continue their study with complementary Masters such as innovative design, management or numerical methods in mathematics, with jointly accredited diplomas such as "Maths in Action: from concept to innovation". A fair proportion of students go on with PhD studies, whether within one of the six research laboratories present inside the school, or in a joint research laboratory where most of the teachers perform their research. In mathematics, all researchers (8) belong to the ICJ laboratory.

6.2.5.6 Curriculum Details

The two first preparatory years in UCBL are representative of the amount of mathematics followed by students. We chose to detail the Electrical Engineering stream in INSA Lyon.

S1:

Algebra 1: Foundations of logic, ensembles, maps, arithmetics, complex numbers, $\mathbb{R}^2$.

Calculus 1: Reals, real functions, sequences, limits, derivation, minimization, maximization, inf, sup, derivation of an implicit function, higher order derivation, convexity, l'Hôpital law, differential equations of first order, primitive.

S2:

Algebra 2: Linear algebra, polynomials, rational fractions, vector spaces of finite and countable dim, linear applications, matrices, determinant.

Calculus 2: Integration and approximation, change of variable, simple elements, circular and hyperbolic formulas, differential equations of second order, indefinite integrals, applications in probability and statistics, Landau notation, limits, Taylor polynomials and series, Taylor–Lagrange remainder.

S3:

Algebra 3: Diagonalization, groups, determinants, eigenspaces, spectral decompositions, Cayley–Hamilton, powers, exponential of a matrix.

Calculus 3: Several variables functions, differential calculus, applications, convexity, Lagrange multipliers, implicit functions, Euler equations, isoperimetric problems.

S4:

Algebra 4: Geometric algebra, bilinear forms, scalar products, rank, kernel, Gauss orthogonalization, adjoint, spectral decomposition of self-adjoint operators, quadratic forms, Sylvester theorem, affine geometry, conics, quadrics, O(p,q).

Calculus 4: Series and sequences, Cauchy, d'Alembert, uniform convergence, Abel theorem, trigonometric and Fourier series, entire series, integrals depending on a parameter, differentiation and continuity, eulerian functions, Laplace transform, applications to differential equations, geometry and differential calculus, curves and surfaces, geometry, parametric curves, curvature, Frenet frame, tangent and normal spaces, vector spaces, differentials, line, surface and multi-dimensional integrals, Stokes theorem, Green formula.

Applied algebra: Gröbner basis, Perron–Frobenius theorem and web indexation.

Student project.

S5:

Algebra 5: Groups and morphisms, Lagrange theorem, cyclic groups, morphism, image, kernel, Euler index, Z/nZ ring, prime numbers, quotient groups, dihedral groups, group action, orbits, stabilizer, Bernside, Sylow theorems, SO_3 subgroups, Platonian solids.

Numerical analysis: Linear algebra, Gauss method, iterative methods, conditioning, spectral problems, power method. Nonlinear equations, Newton, secant method, remainder estimation. Interpolation, approximation, polynomial interpolation, mean squared, numerical integration, discrete Fourier transform, Cooley–Tuckey fast Fourier transform, differential equations, Cauchy problem numerical solution, Euler method, Runge–Kutta, implicit and explicit methods.

Topology: Metric space, normed space, topological space, continuity, Baire lemma, Banach fixed point, Bolzano–Weierstrass theorem, Ascoli theorem, Stone–

Weierstrass theorem, Lebesgue integral, Riemann vs. Lebesgue, Lebesgue dominated convergence theorem, Fubini theorem, Lebesgue measure on $\mathbb{R}^n$, random variables, measures.

Starting with the third year, the curriculum depends on which program we are talking about: INSA, Polytech, CPE, Centrale Lyon. We chose to detail here the Electrical Engineering program of INSA. The amount of mathematics is decreasing with time; most of the technical prerequisites are put in the first year of major study:

S6: $96 + 88 = 184$ h of math for 8 ECTS. Hilbert spaces, Fourier series, distributions, interpolation, approximation, numerical integration, solving linear systems. Fourier, Laplace, Z-transforms, ODEs, linear systems, convolution, difference equations using Matlab.

S7: 83 h for 4 ECTS. Complex analysis, PDE, probabilities. Numerical Analysis: Non linear equations, PDEs Using Matlab.

S8: 60 h 3ECTS. Probability and statistics, conditional probabilities, statistics, moments, central limit theorem, estimation, hypothesis tests, regressions, experimentation plans.

From then on, mathematics is only used as a tool integrated in other teaching units.

The peculiar fact about the preparatory school in UCBL is that, unlike most other preparatory schools, it is organized by units that delivers ECTS: Each semester weights 30 ECTS which come from five teaching units of 6 ECTS. The two first years of the curriculum are followed by all students going to engineer schools at UCBL, but there are some other non-math options shaping them in three different groups: physics and math, computer science and math, and mechanics and math.[6]

---

[6]More details can be found (in French) on http://licence-math.univ-lyon1.fr/doku.php?id=mathgeneappli#premiere_annee_du_parcours; see also http://licence-math.univ-lyon1.fr/doku.php?id=mathgenappliprog. The details for INSA can be found at the following URL: http://www.insa-lyon.fr/formation/offre-de-formation2/g/d/?gr=ING.

Chapter 7
Case Studies of Math Education for STEM in Russia

7.1 Analysis of Mathematical Courses in KNRTU-KAI

Ildar Galeev (✉) and Svetlana Novikova and Svetlana Medvedeva
Kazan National Research Technical University named after A.N. Tupolev—KAI
(KNRTU-KAI), Automated Systems of Information Processing and Control Department, Kazan, Russia
e-mail: monap@kstu.ru; sweta72@bk.ru; pmisvet@yandex.ru

7.1.1 Kazan National Research Technical University named after A.N. Tupolev—KAI (KNRTU-KAI)

Kazan National Research Technical University named after A.N. Tupolev—KAI (KNRTU-KAI) was established in 1932. The history of the University is closely related to the progress of Russian aeronautics. Fundamental education and profound scientific research are the distinguishing features of the university, which make it very attractive for a great number of young people. Until recent times, it was known as Kazan Aviation Institute (KAI).

In 1973, the Institute was named after Andrey N. Tupolev, the prominent aircraft designer. In 1992, it obtained the status of State Technical University. In 2009, KNRTU-KAI became 1 of 12 universities selected among all the Russian universities (from about 900 state HE institutions and 2000 private ones) which was awarded the prestigious title of the "National Research University".

KNRTU-KAI is a member of the European Universities Association EUA (2008) and of the European Association of Aerospace Universities PEGASUS (2009).

© The Author(s) 2018
S. Pohjolainen et al. (eds.), *Modern Mathematics Education for Engineering Curricula in Europe*, https://doi.org/10.1007/978-3-319-71416-5_7

KNRTU-KAI is the largest multidisciplinary educational and scientific complex of the Republic of Tatarstan and the Volga region. The structure KNRTU-KAI includes five institutes, one faculty, five branches in the Republic of Tatarstan (Almetyevsk, Zelenodolsk, Naberezhnye Chelny, Chistopol, Leninogorsk). Despite having the institute of Economics, Management and Social Technologies in its structure, KNRTU should be considered as a technical university. The other four of five main institutes and one faculty are completely devoted to engineering and computer science specializations.

Today KNRTU-KAI is one of the leading Russian institutions in aircraft engineering, engine- and instrument-production, computer science and radio- and telecommunications engineering.

The university includes the following institutes and faculties:

- Institute of Aviation, Land Vehicles & Energetics
- Institute of Automation & Electronic Instrument-Making
- Institute of Technical Cybernetics & Informatics
- Institute of Radio-Engineering & Telecommunications
- Institute of Economics, Management and Social Technologies
- Physics & Mathematics Faculty

There is also the German–Russian Institute of Advanced Technologies (GRIAT). German Partners of this institute are Ilmenau Institute of Technology, TU Ilmenau, Otto-von-Guericke-Universität Magdeburg, OVGU, and Deutscher Akademischer Austauschdienst, DAAD.

Currently there are more than 12,000 students in the university. Most of them (more than 10,000) are technical (STEM) students. In total at six institutes and one faculty there are 29 Bachelor and 30 Master STEM programs (including 9 Master STEM programs of the Institute GRIAT), also there is six Specialist (specific Russian 5-year grade).

The mathematical training at the University is carried out by Department of Higher Mathematics, Department of Special Mathematics, and Department of Applied Mathematics and Computer Science.

Departments of Higher Mathematics (the total number of teachers is 8) and Special Mathematics (the total number of teachers is 18) carry out classical mathematical training in classical engineering programs.

The department of Applied Mathematics and Computer Science trains Bachelors and Masters in the following study programs:

- "Informatics and Computer Science" (B.Sc., MSc);
- "Mathematics and Computer Science" (B.Sc., MSc);
- "Software Engineering" (B.Sc., MSc).

It also provides specialized courses for other areas within the university and in the Institute of Computer Technology and Information Protection. The educational process is realized by the department under the innovation program "Industrial production of software and information technology tools" in accordance with state educational standards of the Russian Federation and international standards on

Computing Curricula and on professional standards in the field of information technology. The staff of this department consists of 6 full-time professors, 12 full-time associate professors, 2 full-time assistant professors, 2 teachers and 6 technical specialists.

There are most general courses in all areas of training; the Institute of Computer Technology and Information Protection provides courses in mathematics: algebra and geometry, calculus, discrete mathematics, probability theory and mathematical statistics, computational mathematics.

The Institute of Computer Technology and Information Protection has six chairs:

- Department of Applied Mathematics and Computer Science
- Department of Information Security Systems
- Department of Computer Aided Design
- Department of Automated Data Processing Systems and Management
- Department of Computer Systems
- Department of Process Dynamics and Control

In 2015, unified training Bachelor's plans were accepted at the university, in which a block of mathematical courses is the same for all STEM programs.

The block of mandatory mathematics for all STEM programs includes the following courses: Calculus, Algebra and Geometry, Probability Theory and Mathematical Statistics (semesters 1–4). The courses Discrete Mathematics and Mathematical Logic and Theory of Algorithms are included in the block of mandatory courses for the training of IT-professionals. The course of Computational Mathematics (semester 5) is in the block of selective courses for the training of IT-professionals. Other mathematical courses, such as Differential Equations, Mathematical Physics, Complex Analysis and Functional Analysis in the selective block, are courses for training of specialists by an in-depth study of mathematics. An example of this Bachelor program is the program of Mathematics and Computer Science.

7.1.2 Comparative Analysis of "Probability Theory and Mathematical Statistics"

"Probability Theory and Mathematical Statistics" is a theoretical course with approximately 3000 students. There are around 240 second year students, from 6 Bachelor and 2 Specialist of Institute of Computer Technology and Information Protection (ICTIP) programs, who study this course at the Department of Applied Mathematics and Computer Science. A comparison of this course was conducted with the corresponding courses "Probability Calculus" and "Statistics" by Tampere University of Technology (TUT). The course outlines are presented in Table 7.1.

Table 7.1 Outlines of probability theory and statistics courses at KNRTU-KAI and TUT

Course information	KNRTU-KAI	TUT
Bachelor/master level	Bachelor	Bachelor
Preferred year	2	2
Selective/mandatory	Mandatory	Mandatory
Number of credits	6	4 + 4
Teaching hours	90	84
Preparatory hours	108	132
Teaching assistants	1	1–4
Computer labs	Available	Available
Average number of students on the course	60	200
Average pass %	85%	90%
% of international students	None	None

"Probability Theory and Mathematical Statistics" is a mandatory Bachelor level course on second year—Probability Theory (third semester) and Mathematical Statistics (4th semester). Its prerequisite courses are: Calculus (first, second semesters), Algebra and Geometry (second semester), Discrete Mathematics (second, third semesters), and Mathematical Logics and Theory of Algorithms (third semester). Prerequisite courses at TUT are Engineering Mathematics 1–4. The follow-up courses are Theory of Stochastic Processes and System, Programming and various special courses of all six ICTIP Bachelor programs.

In 2011, KNRTU-KAI has started to use LMS Blackboard, which is a learning management system supporting e-learning. The content of educational materials is included in the LMS Blackboard in a form suitable for e-learning. For Probability Theory and Mathematical Statistics an e-learning course was developed for the LMS Blackboard environment. It contains the mandatory e-learning components—course schedule, lecture notes, guidelines for tutorials, computer labs, as well as independent student work. There is also a mandatory component—a test for students on theoretical items of the course. Thus, KNRTU-KAI uses a blended form of learning types that combines classroom instruction with students' independent work using the LMS Blackboard environment.

The size of the course is 6 credits, which means on average 216 h of student's work (36 h for each credit). The credits are divided among different activities as follows: lectures 36 h, tutorials 54 h, computer labs 18 h, homework 72 h and 36 h for preparing to the examination.

There are about 240 students studying the course every year at the Institute of Computer Technology and Information Protection. About 10% of them are foreign students and about 20% are female.

The lectures are theoretical, but application examples for every theorem and algorithm are shown as well. Tutorial classes are completely devoted to problem solving, generally using paper and pen, but also MATLAB and Excel are used in solving some problems. Computer labs use our own computer textbook "Introduction to Mathematical Statistics" in real-time. This allows the students to generate a sample from a given distribution, to build and explore random functions and their density distributions, as well as to build and explore their numerical characteristics, and to carry out and explore the algorithms for testing statistical hypothesis and one-dimensional regression analysis.

Generally, our students have to pass four tests and complete four individual laboratory workshops during the two semesters. When all this work has successfully been done, they are allowed to take an exam. In the exam a student has to answer thoroughly two questions from random topics, and to briefly answer some additional questions. Prior to this, all the students of a group had to take a pen and paper test of 15 test items, which allows the teacher to determine the readiness of students for the exam. The final grade is determined by the examiner. The grade depends on how successfully all the parts of the exams were passed, and it takes into account the test results, students' practical work and laboratory work during the semester. The final grade is mathematically dependent on the number of points in accordance with the score-rating system, adopted by the university. The score-rating systems is the following:

- less than 50 points is unsatisfactory (grade "2"),
- from 50 to 69 points is satisfactory (grade "3"),
- from 70 to 84 points is good (grade "4"),
- more than 84 points is excellent (grade "5").

The course is supported by the following educational software and TEL tools:

- MATLAB and Excel are used in tutorials for solving some problems.
- The computer textbook "Introduction to Mathematical Statistics" is used for construction and study of statistical estimations of distributions and their parameters, and also for testing the knowledge and monitoring and evaluating the skills of students in mathematical statistics.
- The e-Learning Systems LMS Blackboard platform is used for testing the students' knowledge on our course.

Since September 2016, we have used the Math-Bridge system for training in and monitoring of students' abilities in solving problems in the theory of probability. The piloting operation of the Math-Bridge system has been done as a part of the MetaMath project.

7.1.2.1   Contents of the Course

The comparison is based on the SEFI framework [1]. Prerequisite competencies are presented in Table 7.2. Outcome competencies are given in Tables 7.3, 7.4, and 7.5.

Table 7.2 Core 0 level prerequisite competencies of probability theory and statistics courses at KNRTU-KAI and TUT

Core 0		
Competency	KNRTU-KAI	TUT
Data handling	With some exceptions[a]	X
Probability	With some exceptions[b]	X
Arithmetic of real numbers	X	X
Algebraic expressions and formulas	X	X
Functions and their inverses	X	X
Sequences, series, binomial expansions	X	X
Logarithmic and exponential functions	X	X
Indefinite integration	X	X
Definite integration, applications to areas and volumes	X	X
Proof	X	X
Sets	X	X

[a]Interpret data presented in the form of line diagrams, bar charts, pie charts; interpret data presented in the form of stem and leaf diagrams, box plots, histograms; construct line diagrams, bar charts, pie charts, stem and leaf diagrams, box plots, histograms for suitable data sets; calculate the mode, median and mean for a set of data items
[b]Define the terms "outcome", "event" and "probability"; calculate the probability of an event by counting outcomes; calculate the probability of the complement of an event; calculate the probability of the union of two mutually exclusive events; calculate the probability of the union of two events; calculate the probability of the intersection of two independent events

Table 7.3 Core 0 level outcome competencies of probability theory and statistics courses at KNRTU-KAI and TUT

Core 0		
Competency	KNRTU-KAI	TUT
Calculate the mode, median and mean for a set of data items	X	X
Define the terms 'outcome', 'event' and 'probability'	X	X
Calculate the probability of an event by counting outcomes	X	X
Calculate the probability of the complement of an event	X	X
Calculate the probability of the union of two mutually exclusive events	X	X
Calculate the probability of the union of two events	X	X
Calculate the probability of the intersection of two independent events	X	X

7.1.2.2 Summary of the Results

The comparison shows that the two courses cover generally the same topics and competences. A difference was observed in the place of the course in the curriculum of the degree program. The course on Probability theory and mathematical statistics is studied in the third and fourth semesters in KNRTU-KAI, but at TUT this course is studied during the fourth and fifth semesters, depending on the given study program. Before MetaMath project there were more differences: in KNRTU-KAI this course was studied in the second semester (first year of training). No difference was observed in the total number of hours.

Table 7.4 Core 1 level outcome competencies of the probability theory and statistics courses at KNRTU-KAI and TUT

Core 1		
Competency	KNRTU-KAI	TUT
Calculate the range, inter-quartile range, variance and standard deviation for a set of data items	X	X
Distinguish between a population and a sample	X	X
Know the difference between the characteristic values (moments) of a population and of a sample	X	X
Construct a suitable frequency distribution from a data set	X	X
Calculate relative frequencies	X	X
Calculate measures of average and dispersion for a grouped set of data	X	X
Understand the effect of grouping on these measures	X	X
Use the multiplication principle for combinations	X	X
Interpret probability as a degree of belief	X	X
Understand the distinction between a priori and a posteriori probabilities	X	X
Use a tree diagram to calculate probabilities	X	X
Know what conditional probability is and be able to use it (Bayes' theorem)	X	X
Calculate probabilities for series and parallel connections	X	X
Define a random variable and a discrete probability distribution	X	X
State the criteria for a binomial model and define its parameters	X	X
Calculate probabilities for a binomial model	X	X
State the criteria for a Poisson model and define its parameters	X	X
Calculate probabilities for a Poisson model	X	X
State the expected value and variance for each of these models	X	X
Understand what a random variable is continuous	X	X
Explain the way in which probability calculations are carried out in the continuous case	X	X
Relate the general normal distribution to the standardized normal distribution	X	X
Define a random sample	X	X
Know what a sampling distribution is	X	X
Understand the term 'mean squared error' of an estimate	X	X
Understand the term 'unbiasedness' of an estimate	X	X

Thus, as a result of the modernization, the number of hours in the course "Probability theory and mathematical statistics" now fully coincide: in the third semester 108 h (3 credit units in the second year) and "Mathematical Statistics" 108 h in the fourth semester (3 credit units in the third year). Studying the course has been shifted from the first to the second year.

A comparison on the use of information technology in the course "Probability Theory and Mathematical Statistics" was carried out in the two universities. The use of information technology in teaching this course is on a high level.

Table 7.5 Core 2 level outcome competencies of the probability theory and statistics courses at KNRTU-KAI and TUT

Core 2		
Competency	KNRTU-KAI	TUT
Compare empirical and theoretical distributions	X	X
Apply the exponential distribution to simple problems	X	X
Apply the normal distribution to simple problems	X	X
Apply the gamma distribution to simple problems	X	X
Understand the concept of a joint distribution	X	X
Understand the terms 'joint density function', 'marginal distribution functions'	X	X
Define independence of two random variables	X	X
Solve problems involving linear combinations of random variables	X	X
Determine the covariance of two random variables	X	X
Determine the correlation of two random variables	X	X
Realize that the normal distribution is not reliable when used with small samples	X	X
Use tables of the t-distribution	X	X
Use tables of the F-distribution	X	X
Use the method of pairing where appropriate	X	X
Use tables for chi-squared distributions	X	X
Decide on the number of degrees of freedom appropriate to a particular problem	X	X
Use the chi-square distribution in tests of independence (contingency tables)	X	X
Use the chi-square distribution in tests of goodness of fit	X	X
Set up the information for a one-way analysis of variance	X	X
Derive the equation of the line of best fit to a set of data pairs	X	X
Calculate the correlation coefficient	X	X
Place confidence intervals around the estimates of slope and intercept	X	X
Place confidence intervals around values estimated from the regression line	X	X
Carry out an analysis of variance to test goodness of fit of the regression line	X	X
Interpret the results of the tests in terms of the original data	X	X
Describe the relationship between linear regression and least squares fitting	X	X
Understand the ideas involved in a multiple regression analysis	X	X
Appreciate the importance of experimental design	X	X
Recognize simple statistical designs	X	X

Table 7.6 Outlines of optimization courses at KNRTU-KAI and TUT

Course information	KNRTU-KAI	TUT
Bachelor/master level	Master	Master
Preferred year	2	2
Selective/mandatory	Mandatory	Selective
Number of credits	3	5
Teaching hours	43	50
Preparatory hours	65	60
Teaching assistants	1	1
Computer labs	Available	Available
Average number of students on the course	20	30
Average pass %	95%	90%
% of international students	–	60%

7.1.3   Comparative Analysis on "Optimisation Methods"

"Optimisation Methods" is a MSc level theoretical course with approximately 20 students (1 academic group). Most of them (80%) are young men. Their ages vary from 20 to 28 years. Most of them are 21 or 22 years old. The course is studied as the second Master's course in the fall semester (third semester). A special feature is that these students have no education in IT—Informatics as their second competence. Teaching the course is carried out at the Department of Applied Mathematics and Computer Science at the Institute of Computer Technology and Information Protection (CTIP). Comparison of this course was conducted with a similar course "Optimisation Methods" by TUT. Outlines of both the courses are presented in Table 7.6.

Optimisation Methods is a mandatory Master's course in the second year of study of the Master's. Prerequisite courses for the Optimisation Methods are: Calculus 1–3, Linear Algebra, Probability theory and Mathematical Statistics, Graph Theory. These courses are part of the Bachelor's degree.

Since 2011, KNRTU-KAI has started to use LMS Blackboard, which is a learning management system supporting e-learning. The content of educational materials are included in the LMS Blackboard in a form suitable for e-learning.

An e-learning course on Optimisation Methods was developed in the LMS Blackboard environment. Mandatory e-learning components—course schedule, lecture notes, guidelines for tutorials, Computer labs, as well as independent student work. There is also a mandatory component—a test for students on theoretical items of the course. Thus, KNRTU-KAI uses a blended form of learning that combines classroom instruction with students' independent work using the LMS Blackboard environment.

The size of the course is 3 credits, which means on average 108 h of work (36 h for each credit). The credits are divided among different activities as follows: lectures 10 h, computer labs 20 h, homework 35 h, 30 h for exam preparation and 3 h for exam.

About 20 students study this course every year at the Institute of Computer Technology and Information Protection. About 20% of them are female.

The lectures are theoretically based, but application examples on all the methods and algorithms are shown also, as well. Computer labs use our own computer tutorial "Optimisation methods", which exists on LMS Blackboard, MS Excel and MATLAB are also in use.

Generally, students have to pass one test and complete five individual laboratory workshops during the semesters. When all this work is successfully done, they are allowed to enter an exam. In the exam a student has to answer thoroughly to two questions from random topics, and briefly to answer some additional questions. Prior to this, all the students of a group had to take a pen and paper test of 40 test items, which allows the teacher to determine the readiness of the students for the exam. The final grade is determined by the examiner. The grade depends on how successfully all the parts of the exams were passed, and it takes into account the test results, students' practical work and laboratory work during the semester. The final grade is mathematically dependent on the number of points in accordance with the score-rating system, adopted by the university.

7.1.3.1 Contents of the Course

The comparison is based on the SEFI framework [1]. Prerequisite competencies are presented in Table 7.7. Outcome competencies are given in Tables 7.8 and 7.9.

Table 7.7 Core 0 level prerequisite competencies of the courses on optimization at KNRTU-KAI and TUT

Core 0		
Competency	KNRTU-KAI	TUT
Arithmetic of real numbers	X	X
Algebraic expressions and formulas	X	X
Linear laws	X	X
Quadratics, cubics, polynomials	X	X
Functions and their inverses	X	X
Logarithmic and exponential functions	X	X
Rates of change and differentiation	X	X
Stationary points, maximum and minimum values	X	X
Definite integration, applications to areas and volumes	X	X
Proof	X	X
Data handling	X	
Probability	X	

Table 7.8 Core 1 level outcome competencies of the courses on optimization at KNRTU-KAI and TUT

Core 1		
Competency	KNRTU-KAI	TUT
Rational functions	X	
Hyperbolic functions	X	
Functions	With some exceptions[a]	X
Differentiation	X	X
Solution of nonlinear equations	X	X
Vector algebra and applications	X	X
Matrices and determinants	X	X

[a]Obtain the first partial derivatives of simple functions of several variables; use appropriate software to produce 3D plots and/or contour maps

Table 7.9 Core 2 level outcome competencies of the courses on optimization at KNRTU-KAI and TUT

Core 2		
Competency	KNRTU-KAI	TUT
Ordinary differential equations	X	
Functions of several variables	With some exceptions[a]	X

[a]Define a stationary point of a function of several variables; define local maximum, local minimum and saddle point for a function of two variables; locate the stationary points of a function of several variables

Unfortunately SEFI Framework does not describe learning outcomes suitable for "Optimisation Methods". After successful completion of the course at KNRTU-KAI, a student should:

- Have experience in using optimization techniques on a PC with Microsoft Windows operating system.
- Be able to formulate basic mathematical optimization problems, depending on the type of quality criteria and availability limitations.
- Be able to solve the problem by the classical method of unconstrained minimization of functions with one and several variables.
- Be able to use the simplex method of linear programming to solve related problems;
- Be able to apply basic numerical methods for solving nonlinear programming in practice.
- Know the fundamentals of modern technologies to develop mathematical models and find optimal solutions.
- Know the classification of extreme problems and methods for solving them.
- Know the main analytical and numerical methods for unconstrained minimization of functions of one and several variables.
- Know the basic methods for solving nonlinear programming problems.
- Know the main algorithms for solving linear programming problems.

7.1.3.2   Summary of the Results

The comparison has shown the following similarities and differences: In both the universities we use blended learning, however, at TUT significantly more hours are devoted to lectures.

The content of the courses in the universities is very similar: the students acquire knowledge on the main topics of conditional and unconditional optimization and linear programming. In Kazan University the dimensional optimization problem is studied separately.

Concerning the use of computer technology. Both universities use learning management systems for presentation of the lecture material, in TUT is Moodle is used, and KNRTU-KAI uses LMS Blackboard. Both systems provide similar opportunities for students. In both universities laboratory work is conducted in specialized programs: in TUT MATLAB is used, in Kazan a specially designed program. The course "Optimisation methods" does not require extensive modernization. To improve the course the best way is to use Math-Bridge technology.

7.1.4   Comparative Analysis on "Discrete Mathematics"

The course "Discrete Mathematics" is designed for Bachelor program "Software Engineering", training profile "Development of software and information systems". The course "Discrete Mathematics" is taught in semesters 2 and 3. In the current academic year 25 students are enrolled on this program on the 2 semester (19 males and 6 females) and 17 students on the 3 semester (11 males and 6 females). The course has been modified to meet the requirements of professional standards of SEFI [1] and implemented in e-learning system Math-Bridge. Modification and development of the e-learning course was carrie out in DFKI by employees group of KNRTU-KAI, who were specially sent there for training and for e-learning course development.

This course was compared with two courses "Discrete Mathematics" and "Algorithm Mathematics" at Tampere University of Technology (TUT), Tampere, Finland. The course outlines can be found in Table 7.10.

Discrete Mathematics is mandatory course for Bachelors of the first and second year of study (2 and 3 semesters). Prerequisite courses for the Discrete Mathematics are: Calculus 1–3, Linear Algebra. Since 2011, KNRTU-KAI started to use learning management system LMS Blackboard. The course material for the e-learning courses are in the LMS Blackboard environment. For Discrete Mathematics an e-learning course was developed into the LMS Blackboard environment. Mandatory e-learning components—course schedule, lecture notes, guidelines for tutorials, computer labs, as well as for independent work of students. There is also a mandatory component—test items for students on the theoretical material of course. Thus, KNRTU-KAI uses a blended form of education that combines classroom instruction with independent work of students in LMS Blackboard.

Table 7.10 Outlines of discrete and algorithm mathematics courses at KNRTU-KAI and TUT

Course information	KNRTU-KAI	TUT
Bachelor/master level	Bachelor	Bachelor
Preferred year	1 and 2	2
Selective/mandatory	Mandatory	Mandatory
Number of credits	12	4+4
Teaching hours	180	49+42
Preparatory hours	252	65+65
Teaching assistants	1	1–3
Computer labs	No	
Average number of students on the course	50	150
Average pass %	85%	85%
% of international students	10%	

The course size is 12 credits, which means on average 432 h of work (36 h for each credit). The credits are divided among different activities as follows: lectures 108 h, tutorials 72 h, homework 252 h, and 72 h for exam preparation.

About 50 students are learning for these course every year at the Institute of Computer Technology and Information Protection. About 20% of them are female.

The lectures are theoretically based, but application examples of every method and algorithm are shown as well. Computer labs use LMS Blackboard, MS Excel and MATLAB.

Generally our students have to pass 4 tests and execute 36 tutorials during the two semesters. When all this work is successfully done, they are allowed to pass an exam. The students have to answer in detail two questions from random topics, and briefly answer some additional questions. Prior to this, all the students have written answers to a test of 15 test items, which allows one to determine the readiness of a student for the exam. The final grade is determined by the examiner depending on how successfully all the parts of the exams were passed. It takes into account the results of the tests, practical work and laboratory work during the semester. The final grade is mathematically dependent on the number of points in accordance with the score-rating system, adopted by the university. The score-rating system is the following:

- less than 50 points is unsatisfactory (grade "2"),
- from 50 to 69 points is satisfactory (grade "3"),
- from 70 to 84 is good (grade "4"),
- more than 84 is excellent (grade "5").

Our course is supported by the following educational software and TEL tools: MATLAB and Excel are used in tutorials to solving some problems and e-Learning Systems LMS Blackboard platform for testing the students' knowledge on our course.

In the next academic year (from fall semester 2017) we will use the international intellectual Math-Bridge system for training and monitoring abilities to solve problems of Discrete Mathematics in the trial operation of the system in accordance with the project MetaMath.

7.1.4.1   Contents of the Course

The comparison is based on the SEFI framework [1]. Prerequisite competencies are presented in Table 7.11. Outcome competencies are given in Tables 7.12, 7.13, 7.14, and 7.15.

Table 7.11 Core 0 level prerequisite competencies of the discrete and algorithm mathematics courses at KNRTU-KAI and TUT

Core 0		
Competency	KNRTU-KAI	TUT
Arithmetic of real numbers	X	X
Algebraic expressions and formulas	X	X
Linear laws	X	X
Quadratics, cubics, polynomials	X	X
Functions and their inverses	X	X
Logarithmic and exponential functions	X	X
Rates of change and differentiation	X	X
Stationary points, maximum and minimum values	X	X
Definite integration, applications to areas and volumes	X	X
Proof	X	X
Data handling	X	X
Probability	X	–

Table 7.12 Core 0 level outcome competencies of the discrete and algorithm mathematics courses at KNRTU-KAI and TUT

Core 0		
Competency	KNRTU-KAI	TUT
Sets	X	X

Table 7.13 Core 1 level outcome competencies of the discrete and algorithm mathematics courses at KNRTU-KAI and TUT

Core 1		
Competency	KNRTU-KAI	TUT
Mathematical logic	X	X
Sets	With some exceptions[a]	X
Mathematical induction and recursion	X	X
Graphs	X	X
Combinatorics	X	

[a]Compare the algebra of switching circuits to that of set algebra and logical connectives; analyze simple logic circuits comprising AND, OR, NAND, NOR and EXCLUSIVE OR gates

Table 7.14 Core 2 level outcome competencies of the discrete and algorithm mathematics courses at KNRTU-KAI and TUT

Core 2		
Competency	KNRTU-KAI	TUT
Number system	With some exceptions[a]	X
Algebraic operators	X	
Recursion and difference equations	With some exceptions[b]	X
Relations	X	X
Graphs	X	X
Algorithms	With some exceptions[c]	X

[a]Carry out arithmetic operations in the binary system
[b]Define a sequence by a recursive formula
[c]Understand when an algorithm solves a problem; understand the worst case analysis of an algorithm; understand the notion of an NP-complete problem (as a hardest problem among NP problems)

Table 7.15 Core 3 level outcome competencies of the discrete and algorithm mathematics courses at KNRTU-KAI and TUT

Core 3		
Competency	KNRTU-KAI	TUT
Combinatorics	X	
Graph theory	X	

7.1.4.2   Summary of the Results

In KNRTU-KAI for the study of the course "Discrete Mathematics" four times more hours allotted than at the Tampere University of Technology. This is due to the fact that this item in KNRTU-KAI is studied for two semesters and only one semester in TUT. As a consequence, many of the topics that are covered in KNRTU-KAI not covered in TUT. In the course "Discrete mathematics" a lot of attention in both universities given to sections related to mathematical logical expressions and algebraic structures. This is useful for students who will continue to study subjects such as "Theory of Algorithms" and "Logic". The TUT students are being prepared with the help of the Moodle e-learning environment. Moodle also helps to gather feedback from students after the course. In KNRTU-KAI it is not used. This experience is very useful.

In general, course "Discrete Mathematics" by KNRTU-KAI is rather good and meets the requirements for IT-students teaching. But it will be useful to make the course some more illustrative by using computer-based training systems. As a result the course "Discrete Mathematics" has been modernized and the e-learning training course was developed in Math-Bridge system.

7.2   Analysis of Mathematical Courses in LETI

Mikhail Kuprianov and Iurii Baskakov and Sergey Pozdnyakov and Sergey Ivanov
and Anton Chukhnov and Andrey Kolpakov and Vasiliy Akimushkin
Saint Petersburg State Electrotechnical University (LETI), Saint Petersburg, Russia
e-mail: mskupriyanov@mail.ru; bosk@bk.ru; sg_ivanov@mail.ru

Ilya Posov
Saint Petersburg State Electrotechnical University (LETI), Saint Petersburg, Russia
Saint Petersburg State University (SPbU), Saint Petersburg, Russia

Sergey Rybin
Saint Petersburg State Electrotechnical University (LETI), Saint Petersburg, Russia
ITMO University, Department of Speech Information Systems, Saint Petersburg,
Russia

7.2.1   Saint Petersburg State Electrotechnical University (LETI)

Saint Petersburg State Electrotechnical University (LETI) was founded in 1886 as a technical college of the post and telegraph department. In 1891 it gained the state of an institute. The first elected university director was the Russian-born scientist A.S. Popov, who is one of the people credited for inventing the radio. From its founding LETI was a center of Russian electrical engineering. Many outstanding Russian scientists have worked there over the years. After the October Revolution, in 1918, the institution was renamed Leningrad Electrotechnical Institute (LETI).

In the 1920s, LETI played a significant role in the development of electrification plans of Russia. After the Second World War, intensive developments of new scientific fields were started at the institute: radio electronics and cybernetics, electrification and automatization of industrial equipment, automatics and telemechanics, computer science and optoelectronics.

In 1991 the institute was renamed after the city and became Saint-Petersburg Electrotechnical Institute. However, it kept Lenin's name in its title. In early 1990s the Faculty of Humanities was founded, and the Institute was granted university status and was renamed again into "V.I.Ulyanov (Lenin) Saint-Petersburg State Electrotechnical University". In 1998 it was renamed for the current name.

Despite having a faculty of humanities and faculty of economics and management in its structure, LETI should be considered as a technical university. The other five of seven main faculties are completely devoted to engineering and computer science specializations.

There is also an Open Faculty (OF) for evening and correspondence students for all specializations. Other faculties, which are the Faculty of Military Education and the newly created Faculty of Common Training do not have their own students;

Table 7.16 Outlines of mathematical logics and theory of algorithms courses at LETI (ML&TA) and TUT (AM)

Course information	LETI	TUT
Bachelor/master level	Bachelor	Bachelor
Preferred year	1	2
Selective/mandatory	Both	Both
Number of credits	5	4
Teaching hours	40	49
Preparatory hours	65	65
Teaching assistants	No	1–2
Computer labs	Available	Available
Average number of students on the course	400	150
Average pass %	85%	90%
% of international students	20%	Less than 5%

they teach students from other faculties. International Student Office works with foreign students, including teaching them Russian. Finally, the Faculty of Retraining and Raising the Level of Skills does not work with students at all, it works with university personnel.

Currently there are more than 8200 students in the university. Most of them (more than 7500) are technical (STEM) students. Totally at five faculties there are 17 Bachelor and 13 Master STEM programs, also there is one Specialist (specific Russian 5-year grade) program at FCTI.

Mathematical education in LETI is divided between two departments: Department of Higher Mathematics-1 and -2.

7.2.2 Comparative Analysis of "Mathematical Logics and Theory of Algorithms" (ML&TA)

"Mathematical Logics and Theory of Algorithms" is a more theoretical course. There are around 400 second year students from all 6 Bachelor and 1 Specialist FCTI programs studying this course. All of them are supposed to be applied specializations, but, of course, each program contains a lot of theoretical courses. The course outlines with the corresponding course from Tampere University of Technology (TUT) "Algorithm Mathematics" are given in Table 7.16.

ML&TA is mandatory Bachelor level course of the second year. Its prerequisite courses are Discrete Mathematics; Linear Algebra and Calculus 1–2. The follow-up courses are Programming and various special courses from all seven FCTI Bachelor programs.

The department responsible for the course is the Department of Higher Mathematics-2. Now the University is in the process of restructuring and probably the next year two departments of mathematics will be joined, but for purpose of teaching FCTI students a new Department of Algorithm Mathematics will be founded. It will be responsible for this course and also for the prerequisite course of Discrete Mathematics.

The overall number of credits is 5. In Russia we have 36 h in 1 credit, so the total amount is 180 h for this course. Among them 36 h of lectures, 36 h of tutorials, 72 h of homework and 36 h of exam preparation. There are about 400 students studying the course every year. About 20% of them are foreign students and about 30% are female.

The lectures are theoretically based, but applications of every theorem and algorithm are shown. Tutorial classes are completely devoted to solving problems, but generally with pen and paper, without computers. However, some of the algorithms are to be implemented by students while studying this course, and almost all of these are used later while passing the follow-up courses. We also offer some students an alternative way to pass an exam by creating a computer program.

The course is generally oriented to individual work. However, group work can be episodically introduced as an experiment.

Generally our students have to pass two tests and complete three individual homeworks during the semester. When all this work is successfully done, they are allowed to enter an exam. The examination system depends on the lecturer: oral or written. The oral exam is a classic Russian form of examination where the student is to answer thoroughly two questions from random topics, solve a problem and briefly answer some additional questions.

The final mark is determined by the examiner depending on how successfully all the parts of the exams were passed. The written exam consists of several problems which are to be solved by students. The final mark mathematically depends on the solved problems ratio. In both cases the final mark belongs to the classic Russian system: from 2 (failed) to 5 (excellent).

Our course is supported by the following TEL tools: Problem generators can create a huge amount of variants of one problem using one or few certain template(s) and changing numbers, letters etc. Of course, this is support for teachers activity, not for students. Google sites are used by teacher and students to exchange information. For example, students can submit some homeworks to the teacher using this sites. The TEL systems were not generally used in this course before the MetaMath project. Introducing TEL systems (Moodle and our own subject manipulators) is the main direction of our course modification.

7.2.2.1 Contents of the Course

The comparison is based on the SEFI framework [1]. Prerequisite competencies are presented in Table 7.17. Outcome competencies are given in Tables 7.18, 7.19, 7.20, 7.21, and 7.22.

Table 7.17 Prerequisite competencies of mathematical logics and theory of algorithms courses at LETI (ML&TA) and TUT (AM)

Core 1		
Competency	LETI	TUT
Arithmetic of real numbers	X	X
Algebraic expressions and formulas	X	X
Linear laws	X	X
Quadratics, cubics, polynomials	X	X
Functions and their inverses	X	X
Sequences, series, binomial expansions	X	X
Logarithmic and exponential functions	X	X
Proof	X	X
Sets	X	X

Table 7.18 Core 1 level prerequisite competencies of mathematical logics and theory of algorithms courses at LETI (ML&TA) and TUT (AM)

Core 1		
Competency	LETI	TUT
Sets	With some exceptions[a]	X
Mathematical induction and recursion	X	X
Graphs	X	
Matrices and determinants	X	X
Combinatorics	X	

[a]Excluding logical circuits

Note: sometimes graphs are completely included in Discrete Mathematics, then they should be considered as prerequisites. In other cases they are divided between courses of DM and ML&TA, and then they should be partially considered as outcomes

Table 7.19 Core 2 level prerequisite competencies of mathematical logics and theory of algorithms courses at LETI (ML&TA) and TUT (AM)

Core 2		
Competency	LETI	TUT
Number systems	X	
Algebraic operations	Excluding hamming code	X
Relations	Excluding inverse binary relations and ternary relations	X

Note: Binary relations sometimes are included in Discrete Mathematics, so they are considered as prerequisites; in other cases they are included in ML&TA and so they are outcome competences

Table 7.20 Core 1 level outcome competencies of mathematical logics and theory of algorithms courses at LETI (ML&TA) and TUT (AM)

Core 1		
Competency	LETI	TUT
Mathematical logic	X	X
Graphs	X	

Table 7.21 Core 2 level outcome competencies of mathematical logics and theory of algorithms courses at LETI (ML&TA) and TUT (AM)

Core 2		
Competency	LETI	TUT
Relations	Excluding inverse binary relations and ternary relations	X
Algorithms	X	X

Note: Binary relations sometimes are included in Discrete Mathematics, so they are considered as prerequisites; in other cases they are included in ML&TA and so they are outcome competences

Table 7.22 Core 3 level outcome competencies of mathematical logics and theory of algorithms courses at LETI (ML&TA) and TUT (AM)

Core 3		
Competency	LETI	TUT
Find the distance (shortest way) between two vertices in a graph	X	
Find a the graph and his matrix for a relation	X	
Use topological sort algorithm and transitive closure algorithms	X	
Understand the concept of Boolean function	X	X
Construct a truth table for a function	X	X
Obtain CNF and DNF of a function	X	
Obtain Zhegalkin polynomial of a function	X	
Build a composition of two or more functions in different forms	X	
Recognize function membership in one of the post classes	X	
Use post criteria for a set of functions	X	
Recognize context-free grammar	X	
Construct context-free grammar for a simple language	X	
Build a parser for a grammar using Virt algorithm	X	
Recognize table and graph representation of final state machine	X	
Recognize automata language	X	
Carry out set operations with automata languages	X	
Obtain FSM for regular expression and vice versa	X	
Obtain determined FSM for non-determined one	X	
FSM minimization	X	
Understand the notion of a turing machine	X	
Run simple turing machines on paper	X	
Construct simple turing machine	X	
Run Markov algorithm	X	
Recognize the prenex and Skolem form of first-order formulas	X	
Obtain the prenex and Skolem form for a certain formula	X	
Unify first-order logic formulas	X	
Use resolution method for propositions and first-order logic	X	

7.2.2.2 Summary of the Results

The comparison shows that both courses cover generally the same topics and competences. However LETI ML&TA course contains more competences than the corresponding Algorithm Mathematics course in TUT. It follows from two main reasons: first, in LETI the course has more credits and hours; and second, LETI course is more intensive which is good for gifted students and may be probably be not so good for some others. This conclusion does not lead to any course modifications just because we do not want to reduce our course.

The second conclusion is that TUT's course is more applied and uses more TEL systems. There is space for modifications of our course. The main idea of modification is to introduce Moodle in our course and add there as much as possible lectures, test and laboratory works which could help our students to get closer to understanding of our course through their self-activity on the internet.

7.3 Analysis of Mathematical Courses in UNN

Dmitry Balandin and Oleg Kuzenkov and Vladimir Shvetsov
Lobachevsky State University of Nizhny Novgorod (UNN), Nizhny Novgorod, Russia
e-mail: dbalandin@yandex.ru; kuzenkov_o@mail.ru; shvetsov@unn.ru

7.3.1 Lobachevsky State University of Nizhni Novgorod (UNN)

The Lobachevsky State University of Nizhni Novgorod (UNN) is one of the leading classical research universities in Russia established in 1916. The university provides fundamental education in accordance with the best traditions of Russian Higher Education.

By the decision of the Russian Government, in 2009 UNN was awarded the prestigious status of a National Research University.

Being an innovative university, the University of Nizhni Novgorod provides high-quality research-based education in a broad range of academic disciplines and programs. The combination of high educational quality and accessibility of education due to a great variety of educational program types and forms of training is a distinctive feature of the University in today's global knowledge economy.

UNN is ranked 74 by the QS World University Ranking BRICS, and it has five QS Stars for excellence in Teaching, Employability, Innovation, and Facilities. UNN is one of only 15 Russian universities awarded in 2013 with a prestigious grant of the Government of the Russian federation to implement the Leading Universities International Competitiveness Enhancement Program.

Holding leading positions among Russian universities, it is a worldwide recognized institute of higher education: UNN is represented in the Executive Board of the Deans European Academic Network (DEAN), it is a member of the European University Association (EUA) and has direct contractual relations with more than 20 foreign educational and research centers.

The UNN has been implementing a great number of international projects, funded by the European Commission, IREX (the International Research & Exchanges Board), National Fund for Staff Training and other widely known organizations. The University also closely cooperates with the institutes of the Russian Academy of Sciences and the largest transnational high-tech companies (Microsoft, Intel etc.).

At present about 40,000 students and post-graduate students from more than 65 countries of the world are studying at UNN, about half of which pursue STEM courses. Training on all educational programs is conducted by a highly qualified teaching staff, over 70% of them having a PhD or Doctor of Science degrees.

Lobachevsky State University of Nizhni Novgorod as a National Research University has been granted the right to develop its own self-imposed educational standards (SIES). In 2010, the first UNN standard was developed in the area of studies "010300 Fundamental Computer Science and Information Technology (FCSIT)" (Bachelor's degree). In 2011, the second UNN standard was developed in the area of studies "Applied Computer Science" (ACS) (Bachelor's degree).

The Bachelor's program "Information Technologies", is aimed at training experts in high-level programming for hi-tech companies of the information industry.

UNN is engaged in successful cooperation with major international IT companies like Intel, Microsoft, IBM, Cisco Systems, NVIDIA that provide the University with advanced computer equipment and software. This ensures that the educational process is based on the latest achievements in this field of science and technology. At the University, there are research laboratories established with the support of Intel Corporation as well as educational centers of Microsoft and Cisco Systems. In 2005 Bill Gates, president of Microsoft Corporation, named the University of Nizhni Novgorod among the world's ten leading universities in the field of high-performance computing. In 2013 UNN built the powerful supercomputer "Lobachevsky", which is the second fastest supercomputer in Russia.

Teachers, working in this program, are all recognized experts in various fields of science, Doctors and Candidates of Science. The program of studies in Information Technologies is envisaged by Computing Curricula 2001 recommended by such international organizations as IEEE-CS and ACM.

Graduates of the Bachelor's program "Fundamental Computer Science and Information Technologies", are prepared for the following activities in their professional sphere:

- scientific and research work in the field of theoretical computer sciences, as well as development of new information technologies;
- design and application of new information technologies, realized in the form of systems, products and services;

- application of information technologies in project designing, management and financial activities.

In 2014, UNN educational program in the area of studies "010300 Fundamental Computer Science and Information Technology (FCSIT)" (Bachelor's degree) received the accreditation of Russian Engineer Education Association. As a rule, UNN IT-students study in 3 groups with 20 students in each one. Mean age is 17. Male students are twice as many as females. Moreover, there is one group with 20 foreign students (Asia, Africa), who study in English.

7.3.2   Comparative Analysis of "Mathematical Analysis"

Mathematical Analysis (Calculus) is included in curricula as one of the core subjects of mathematics with their own distinct style of reasoning. Mathematical Analysis is ubiquitous in natural science and engineering, so the course is valuable in conjunction with Engineering majors. The purpose of the course is to provide a familiarity to concepts of the real analysis, such as limit, continuity, differentiation, connectedness, compactness, convergence etc. The number of students is 60. It is a mandatory course in UNN that combines both theoretical and applied approaches. Mathematics plays key role in the course and this course forms a foundation for several other applied special disciplines in corresponding educational programs. Mathematical Analysis contains three parts corresponding to the semesters of study: Mathematical Analysis I, II, and III.

Mathematical Analysis I contains such topics as basic properties of inverse, exponential, logarithmic and trigonometric functions; limit, continuity and derivative of a function, evaluating rules of a derivative, function research and curve-sketching techniques, applications of derivative in the optimization problems, L'Hôpital's rule. Mathematical Analysis II contains such topics as indefinite and definite integral and their properties, rules of integration, integration of rational functions, evaluating area between curves and surface area using integrals, integrals application in physics, functions of several variables, implicit functions. Mathematical Analysis III contains such topics as theory of number series and functional series, Fourier series, double integrals and further multiple integrals, line and surface integrals.

There are three comparisons: first UNN Mathematical Analysis I (MAI) and Tampere University of Technology (TUT) Engineering Mathematics 1 (EM1), second UNN Mathematical Analysis II (MAII) and TUT Engineering Mathematics 3 (EM3), and third UNN Mathematical Analysis III (MAIII) and TUT Engineering mathematics 4 (EM4). The course outlines are seen in Tables 7.23, 7.24, and 7.25, respectively.

There are no prerequisite courses. The follow-up course for Mathematical Analysis is Differential Equations. In UNN this course is included in the group of core mandatory mathematical courses of the corresponding educational programs.

Table 7.23 Outlines of MAI (UNN) and EM1 (TUT) courses

Course information	UNN	TUT
Bachelor/master level	Bachelor	Bachelor
Preferred year	1	1
Selective/mandatory	Mandatory	Mandatory
Number of credits	6	5
Teaching hours	108	57
Preparatory hours	180	76
Teaching assistants	3	1–3
Computer labs	No	Available
Average number of students on the course	72	200
Average pass %	90%	90%
% of international students	25%	Less than 5%

Table 7.24 Outlines of MAII (UNN) and EM3 (TUT) courses

Course information	UNN	TUT
Bachelor/master level	Bachelor	Bachelor
Preferred year	1	1
Selective/mandatory	Mandatory	Mandatory
Number of credits	6	5
Teaching hours	108	62
Preparatory hours	180	76
Teaching assistants	3	1–3
Computer labs	No	Available
Average number of students on the course	72	200
Average pass %	90%	90%
% of international students	25%	Less than 5%

Table 7.25 Outlines of MAIII (UNN) and EM4 (TUT) courses

Course information	UNN	TUT
Bachelor/master level	Bachelor	Bachelor
Preferred year	2	1–2
Selective/mandatory	Mandatory	Selective
Number of credits	8	4
Teaching hours	144	49
Preparatory hours	72	57
Teaching assistants	3	1–2
Computer labs	No	Available
Average number of students on the course	65	150
Average pass %	90%	85%
% of international students	25%	Less than 5%

Teaching the course in UNN is more theory-based and unfortunately does not include any innovative pedagogical methods and tools such as: blended learning, flipped classroom, MOOCs, project-based learning, inquiry-based learning, collaborative learning, etc. In TUT on the other hand one can find blended learning, collaborative learning, project-based approach and active use of modern TEL tools for administration, teaching and assessment purposes. Not all modern pedagogical technologies are used in TUT but those of them that do exist in educational process are applied widely and successfully.

The overall number of hours and credits in Mathematical Analysis I, II, and III is 20 cu. = 720 h. It consists of 202 h lectures, 183 h tutorials and 200 h independent work (homework) and 135 h control of independent work (exams) There are two types of homework assignments: these are problems which arise while lecturing, assigned almost every class day and set of problems assigned during practical lessons (weekly). Tests and exams are conducted some times per each term in written, electronic, and oral forms.

There are the following types of assessment used in UNN: positive (perfect, excellent, very good, good, satisfactory); or negative (unsatisfactory, poor). For perfect, the student displays in-depth knowledge of the main and additional material without any mistakes and errors, can solve non-standard problems, has acquired all the competences (parts of competences) relating to the given subject in a comprehensive manner and above the required level. A stable system of competences has been formed, interrelation with other competences is manifested.

For excellent grade, the student displays in-depth knowledge of the main material without any mistakes and errors, has acquired all the competences (parts of competences) relating to the given subject completely and at a high level, a stable system of competences has been formed. For very good grade, the student has sufficient knowledge of the main material with some minor mistakes, can solve standard problems and has acquired completely all the competences (parts of competences) relating to the given subject. For good grade, the student has the knowledge of the main material with some noticeable mistakes and has acquired in general the competences (parts of competences) relating to the given subject.

For satisfactory grade the student has the knowledge of the minimum material required in the given subject, with a number of errors, can solve main problems, the competences (parts of competences) relating to the subject are at the minimum level required to achieve the main learning objectives. If the grade is unsatisfactory, the knowledge of the material is insufficient, additional training is required, the competences (parts of competences) relating to the subject are at a level that is insufficient to achieve the main learning objectives. Finally, for poor grade, there is lack of knowledge of the material, and relevant competences have not been acquired.

There are two midterm exams (tests) at the end of the first and second semesters and there is a final exam (test) at the end of the third semester.

7.3.2.1   Contents of the Course

The course comparison is based on the SEFI framework [1]. Prerequisite competencies of the MAI and EM1 courses are given in Table 7.26. Core 1, core 1, core 2, and core 3 outcome competencies for MAI and EM1, MAII and EM3, MAIII and EM4, are given in Tables 7.26, 7.27, 7.28, 7.29, 7.30, 7.31, and 7.32.

Table 7.26 Core 0 level prerequisite competencies of the MAI (UNN) and EM1 (TUT) courses

Core 0		
Competency	UNN	TUT
Functions and their inverses	X	X
Progressions, binomial expansions	X	X
Logarithmic and exponential functions	X	X
Rates of change and differentiation	X	X
Maximum and minimum values	X	X
Proof	X	

Table 7.27 Core 1 level outcome competencies of the MAI (UNN) and EM1 (TUT) courses

Core 1		
Competency	UNN	TUT
Hyperbolic functions	X	X
Rational functions	X	X
Functions	X	X
Differentiations	X	X
Sequences and series	Only sequences	In EM3

Table 7.28 Core 0 level outcome competencies of MAII (UNN) and EM3 (TUT) courses

Core 0		
Competency	UNN	TUT
Indefinite integral	X	X
Definite integral, applications	X	X

Table 7.29 Core 1 level outcome competencies of MAII (UNN) and EM3 (TUT) courses

Core 1		
Competency	UNN	TUT
Methods of integration	X	X
Application of integration	X	X

Table 7.30 Core 2 level outcome competencies of MAII (UNN) and EM3 (TUT) courses

Core 2		
Competency	UNN	TUT
Functions of several variables	X	In EM4
Nonlinear optimization	X	
Fourier series	X	

Table 7.31 Core 1 level outcome competencies of MAIII (UNN) and EM4 (TUT) courses

Core 1		
Competency	UNN	TUT
Sequences and series	Only series	In EM3

Table 7.32 Core 2 level outcome competencies of the MAIII (UNN) and EM4 (TUT) courses

Core 2		
Competency	UNN	TUT
Function of several variables		X
Nonlinear optimization		X
Fourier series	X	
Double integrals	X	X
Further multiple integrals	X	X
Vector calculus	X	
Line and surface integrals	X	

7.3.2.2   Summary of the Results

Comparative analysis shows that thematic content and learning outcomes for both universities are quite close. The difference is observed in the number of hours. The total number of hours is about 700 in UNN, whereas in TUT it is about 400. New information technologies and, in particular, e-learning systems are actively used in TUT. This allows TUT to take out some of the material for independent study and focus on really difficult topics of the discipline. E-learning systems also allow one to automate and, as a result, simplify the knowledge assessment process. In UNN these information technologies are occasionally used some times per semester. UNN established pre- and posttest to control incoming and outcoming students' knowledge in mathematical analysis. The electronic course in the Moodle system is implemented for teaching mathematical analysis for students in study programs AMCS and FCSIT (Applied Mathematics and Computer Sciences, Fundamental Computer Sciences and Information Technologies, respectively). All tests are based on SEFI competences; they contain a large amount of simple tasks (during 60 min students must fulfill 20 tasks) that allow one to control 160 SEFI competencies from the zeroth to the second level in the areas of "Analysis and Calculus". Authors used Moodle system rather than Math-Bridge because it is more cross-platform and will help the project results to be more sustainable.

The main steps of the course modernization are: including new bridging section "Elementary Mathematics" at the beginning of Mathematical Analysis I; increasing the number of seminars and decreasing the number of lectures; increasing the number of consultations (from 15 to 30 h); mandatory regular testing students during the term (includes using Math-Bridge) two tests per term; increasing the number of engineering examples in the course; using project learning (two projects per term at least). The topics of the projects are: "Approximate calculation of functions: a creation of the calculator for logarithms, trigonometric and hyperbolic functions", "Technical and physical applications of derivatives", "Research of the

Table 7.33 Outlines of mathematical modeling courses at UNN and TUT

Course information	UNN	TUT
Bachelor/master level	Bachelor	Both
Preferred year	3	3–4
Selective/mandatory	Selective	Selective
Number of credits	3	5
Teaching hours	36	28
Preparatory hours	54	80
Teaching assistants	–	–
Computer labs	Available	Available
Average number of students on the course	10	60
Average pass %	100%	95%
% of international students	0%	Less than 5%

normal distribution, the logistic function, the chain line", "The calculation of the center of gravity", "Applications of Euler integral", and so on.

7.3.3 Comparative Analysis of "Mathematical Modeling"

The subject of study in this course are modeling methods and relevant mathematical model used in a variety of subject areas. As a result students should know methods of mathematical modeling. The course helps to learn how to model situations in order to solve problems. The course is based on the theory of differential equations and the theory of probability. There is a final test at the end of the semester. The "Mathematical Modeling" course at UNN is compared with the similar course "Basic Course on Mathematical Modeling" at Tampere University of Technology (TUT). Course outlines are given in Table 7.33.

The main goal of the course is in studying the fundamental methods of mathematical modeling. It contains such topics as the history of modeling, classes of models, differential equations and systems as mathematical models, dynamic systems, mathematical models in physics, chemistry, biology, ecology, models of a replication, the selection processes, continuous and discrete models of behaviour, models of adaptive behaviour, models of decision making, models of a selection of strategies, the selection and optimization, models of social and economic behaviour, optimal control, information models, the model of transmission and storage of information.

7.3.3.1 Contents of the Course

The comparison is based on the SEFI framework [1]. Prerequisite competencies are presented in Table 7.34.

Table 7.34 Core 2 level prerequisite competencies of the mathematical modeling courses at UNN and TUT

Core 2		
Competency	UNN	TUT
Ordinary differential equations	X	X
The first-order differential equations	X	X
The second-order differential equations	X	X
Eigenvalue problems	X	X
Nonlinear optimization	X	X

7.3.3.2   Summary of the Results

The main steps of the course modernization are decreasing the number of lectures (from 36 h to 18 h), including independent work by students (18 h), mandatory regular testing of students (four times during the term, includes the use of Math-Bridge), using engineering examples in the course, using method of project learning (four projects per term). There are the following projects:

1. Introduction in mathematical modeling (simplest models: Volterra, Verhulst, Laurence, Lotka etc.). The aim of the project is to understand principles of mathematical modeling and the entity of the project's work.
2. Modeling eco-systems. The aim of the project is to apply qualitative research methods for mathematical models given in the form of systems of differential equations.
3. Chemical kinetics modeling. The aim of the project is to apply the Lyapunov function method.
4. The final project. The aim of the project is to apply information technologies in mathematical modeling. The topics of the projects are "The calculation of the index of competitiveness", "The calculation of evolutionary stable daily vertical migrations of aquatic organisms", "Models of strategies for socio-economic behaviour", "Neural network models", and so on.

7.4   Analysis of Mathematical Courses in OMSU

Sergey Fedosin
Ogarev Mordovia State University (OMSU), Department of Automated Systems of Information Management and Control, Saransk, Russia

Ivan Chuchaev
Ogarev Mordovia State University (OMSU), Faculty of Mathematics and IT, Saransk, Russia
e-mail: mathan@math.mrsu.ru

Aleksei Syromiasov
Ogarev Mordovia State University (OMSU), Department of Applied Mathematics, Differential Equations and Theoretical Mechanics, Saransk, Russia
e-mail: syal1@yandex.ru

7.4.1 *Ogarev Mordovia State University (OMSU)*

Ogarev Mordovia State University (hereinafter OMSU) is a classical university, though a wide range of technical major programs are presented here. Since 2011 OMSU has the rank of National Research university, which reflects the university's high status in research in Russia. More than 24,000 students study in OMSU, almost 13,000 of them have intramural instruction.

There are several institutions in OMSU that may be referred to as STEM. They are: faculty of Architecture and Civil Engineering, faculty of Biotechnology and Biology, faculty of Mathematics and IT, institute of Mechanics and Energetics, institute of Physics and Chemistry, institute of Electronics and Lighting Technology and a branch in the town of Ruzaevka (institute of Machine-Building). In these institutions there are about 5000 students.

It is difficult to determine the number of STEM disciplines because every STEM major profile (or group of major profiles) has its special curriculum in mathematics, its special curriculum in physics and so on. So there is variety of STEM disciplines with the same name but with different contents. Totally, there are more than 100 different STEM disciplines in OMSU.

OMSU is the oldest and the biggest higher education institution in the Mordovian Republic. Its main campus is situated in Saransk, which is the capital of Mordovia. The institution was founded at the 1st of October, 1931 as Mordovian Agropedagogical Institute. Next year it was transformed into Mordovian Pedagogical Institute. Based on this institute Mordovian State University was organized in 1957. Finally, in 1970 it was named after the poet N.P. Ogarev.

More than 150,000 people graduated from the university during its history. OMSU graduates formed the backbone of scientific, engineering, pedagogical, medical and administrative staff in the Mordovian Republic. Many of the graduates work in other regions of Russia or abroad.

7.4.2 *Comparative Analysis of "Algebra and Geometry"*

Algebra and Geometry (AlGeo) for the major study programs Informatics and Computer Science (ICS) and Software Engineering (SE) is a fundamental mathematical course, so it is more of a theoretical than an applied course. OMSU tries to make it more applied by giving students programming tasks (for example, students must write a computer program which solves linear systems of equations using Gaussian elimination). Totally, there are about 50 first year students in two study programs (ICS and SE), and all of them must study this course.

The course information is to be compared with our European partner in the project Université Claude Bernard Lyon 1 (UCBL). The course outlines can be seen in Table 7.35.

Table 7.35 Outlines of algebra and geometry courses at OMSU and UCBL

Course information	OMSU	UCBL
Bachelor/master level	Bachelor	Bachelor
Preferred year	1	1
Selective/mandatory	Mandatory	Mandatory
Number of credits	6 (216 h)	6 + 6
Teaching hours	108	60 + 60
Preparatory hours	108	120
Teaching assistants	No	yes
Computer labs	Several programming tasks as homework	Gaussian elimination using Sage-math
Average number of students on the course	50	189
Average pass %	85–90%	65%
% of international students	Less than 10%	14%

A prerequisite for Algebra and Geometry is mathematics as studied at secondary school. Follow-up courses are Calculus, Discrete Mathematics, Probability theory and Statistics. The course of Algebra and Geometry is included in the group of mandatory mathematical courses that all the students in the ICS and IT-study programs must study during the first years.

Though course contents for ICS and SE study programs are almost the same, there are two departments responsible for the course of Algebra and Geometry. The Department of Calculus is responsible for this course for ICS profile. There are two full professors and ten associate professors working in this department. The Department of Applied Mathematics, Differential Equations and Theoretical Mechanics is responsible for this course for SE-study program. Four full professors, 14 associate professors and 3 teachers work in this department.

The overall number of credits for the course is 6. Russian credit is 36 h, so the total amount is 216 h for this course. Among them one has 36 h of lectures, 72 h of tutorials and 108 h of homework.

There is one 2-h lecture on Algebra and Geometry and two 2-h tutorials every week. During the tutorials students solve some problems (fulfill computational tasks) under the teacher's direction and control. Students may be given home tasks, which must be done during preparatory hours. We do not need computer labs for every tutorial, but some home tasks for this course are programming tasks. Students write simple programs on topics of the course, for example, they must write a program that solves a linear equation system. The programming language is of the student's choice. Totally, there are about 25 ICS students and about 25 SE students attending the course; for these profiles, lectures and tutorials are set separately. It is hard to mention the number of students who finish this certain course, because student expulsion in OMSU is the result of failing of two or more courses. But about 10–15% students of SE and ICS profiles usually drop out after the first 1 or 2 years of study, and as a rule, these students have problems in mathematics. Usually only Russian students study on ICS and SE programs; in OMSU most popular for international students is the Medical Institute.

The average age of students attending the course is 18, and about 75% of the students are male. In OMSU there is no mandatory formal procedure of course rating; unofficial feedback contains more likes than dislikes of Algebra and Geometry.

The course of Algebra and Geometry is established for the first year students and is quite theoretical. So the pedagogy is traditional: students listen to lectures, fulfill some tasks during tutorials and do their homework. We think that younger students must have less educational freedom than older ones, and the role of the teacher in the learning process for younger students must be more explicit. That is why we do not use project-based learning in this course. But, of course, we try to make the learning process more interesting, sometimes funny and even competitive. For example, from time to time a group of students in the tutorial is divided into several subgroups and every subgroup fulfills some task. Solving linear equations systems by Cramer's rule is a good example of a task that can be "parallelized" so that every student in the subgroup does his/her part of the total work, and students in a subgroup collaborate. This kind of work in subgroups is very competitive and students like it. Blended learning is used episodically: some teachers use Moodle for distance learning. But for the students who have resident instruction (and here we discuss these students) it is more the exception than the rule.

A rating system is used for assessment at OMSU. The maximum rating is 100 points; one can get 70 points during the semester and only 30 points (as a maximum) is left for the exam procedure. In a semester students get their points for work in the classes and for fulfilling two–three large tests. These tests include a large amount of tasks; not only the answers, but the solutions are controlled by the teacher. An exam is passed in oral form and it includes two theoretical questions (like a theorem with a proof) and a computational task. A student's final rating sums up the semester and exam ratings. If this sum is 86 or more, the student's knowledge of the course is graded as excellent (ECTS grade A or B); if the sum is between 71 and 85, student has grade "good" (approximately ECTS grade C or D); if the sum is between 51 and 70, student's grade is "satisfactory" (ECTS grade E). Finally, a student fails (gets grade "non-satisfactory" which is equivalent to ECTS grade F) if his/her rating is 50 or less.

As for technology, high-level programming languages (C++ or Pascal) are used for homework. Programming (topics are harmonized with the course contents) is a mandatory part of the tests fulfilled by the students during the semester. This programming activity has an influence on the student's final rating.

Until 2014, TEL systems were not used in teaching of the course for students with resident instruction, but after participating in MetaMath project we plan to use Math-Bridge, GeoGebra and, perhaps, Moodle in teaching Algebra and Geometry. Now Moodle is used by students who study distantly (such distant learning is not the part of resident instruction now and is not included in this analysis).

E-mail and social networks are used sometimes to have a closer connection with students, to provide tasks for them and so on.

7.4.2.1   Contents of the Course

The comparison is based on the SEFI framework [1], for each level from Core 0 to Core level 3 only subareas of mathematics are listed. The symbol "X" means that all the competencies from this subarea are prerequisite for the course; exclusions are listed in explicit form. If few competencies (minor part) from the subarea are prerequisite, the description begins with "Only....".

Core 1 and Core 1 level prerequisite competencies are presented in Tables 7.36 and 7.37. Outcome competencies of level zero, one, two and three, are given in Tables 7.38, 7.39, 7.40, and 7.41.

Table 7.36 Core 0 level prerequisite competencies of algebra and geometry courses at OMSU and UCBL

Core 0		
Competency	OMSU	UCBL
Arithmetic of real numbers	X	X
Algebraic expressions and formulas	X	X
Linear laws	With some exceptions[a]	X
Quadratics, cubics and polynomials	X	X
Proof	X	X
Geometry	X	X
Trigonometry	X	X
Coordinate geometry	Only[b]	X
Trigonometric functions and applications	X	X
Trigonometric identities	X	X

[a]Obtain and use the equation of a line with known gradient through a given point; obtain and use the equation of a line through two given points; use the intercept form of the equation of a straight line; use the general equation $ax + by + c = 0$; determine algebraically whether two points lie on the same side of a straight line; recognize when two lines are perpendicular; interpret simultaneous linear inequalities in terms of regions in the plane

[b]Calculate the distance between two points; give simple example of a locus; recognize and interpret the equation of a circle in standard form and state its radius and center; convert the general equation of a circle to standard form

Table 7.37 Core 1 level prerequisite competencies of algebra and geometry courses at OMSU and UCBL

Core level 1		
Competency	OMSU	UCBL
Vector arithmetic	Excluding: determine the unit vector in a specified direction	X
Vector algebra and applications	Excluding: all competencies due to vector product and scalar triple product	X

Table 7.38 Core 0 level outcome competencies of algebra and geometry courses at OMSU and UCBL

Core 0		
Competency	OMSU	UCBL
Linear laws	Those not in prereq.	X
Functions and their inverses	Only[a]	Only[b]
Rates of change and differentiation	Only: obtain the equation of the tangent and normal to the graph of a function	same
Coordinate geometry	Those not in prereq.	X

[a]Understand how a graphical translation can alter a functional description; understand how a reflection in either axis can alter a functional description; understand how a scaling transformation can alter a functional description
[b]All competencies excluding the properties of $1/x$ and the concept of limit (Calculus)

Table 7.39 Core 1 level outcome competencies of algebra and geometry courses at OMSU and UCBL

Core 1		
Competency	OMSU	UCBL
Conic sections	X	X
3D coordinate geometry	X	X
Vector arithmetic	Only: determine the unit vector in a specified direction	X
Vector algebra and applications	Only: all competencies due to vector product and scalar triple product	X
Matrices and determinants	Excluding: use appropriate software to determine inverse matrices	Sage-math used
Solution of simultaneous linear equations	X	X
Linear spaces and transformations	With some exceptions[a]	X

[a]Define a subspace of a linear space and find a basis for it; understand the concept of norm; define a linear transformation between two spaces; define the image space and the null space for the transformation

7.4.2.2   Summary of the Results

The main findings that course comparison has shown are: More exact name for OMSU course of Algebra and Geometry should be "Linear Algebra and Analytic Geometry". The French course is much more fundamental and much more extensive. Though OMSU course of Algebra and Geometry is one of the most theoretical in Software Engineering and Informatics and Computer Science study programs, it is more adapted to the specificity of these programs. It is less fundamental than the French one and this is the price to pay for having numerous IT-courses in the curriculum. Though the UCBL course of Algebra and Geometry is in total much more extensive, the amount of students' learning load per semester is

Table 7.40 Core 2 level outcome competencies of algebra and geometry courses at OMSU and UCBL

Core 2		
Competency	OMSU	UCBL
Linear optimization	Only[a]	Same
Algorithms	Only: understand when an algorithm solves a problem; understand the 'big O' notation for functions	
Helix	Only: recognize the parametric equation of a helix	
Geometric spaces and transformations	With some exceptions[b]	With some exceptions[c]

[a]Recognize a linear programming problem in words and formulate it mathematically; represent the feasible region graphically; solve a maximization problem graphically by superimposing lines of equal profit

[b]Understand the term 'invariants' and 'invariant properties'; understand the group representation of geometric transformations; classify specific groups of geometric transformations with respect to invariants

[c]Apply the Euler transformation, cylindrical coordinates, group representation, classification of group representation

Table 7.41 Core 3 level outcome competencies of algebra and geometry courses at OMSU and UCBL

Core 3		
Competency	OMSU	UCBL
Geometric core of computer graphics	With some exceptions[a]	
Matrix decomposition	Strassen's algorithm for quick multiplying of matrices	

[a]Write a computer program that plots a curve which is described by explicit or parametric equations in cartesian or polar coordinates; know Bresenham's algorithm and Xiaolin Wu's algorithm of drawing lines on the display monitor

bigger in OMSU's course (108 vs. 60 teaching hours, that is, 80% more). In OMSU the percent of teaching hours is more than in UCBL (50% vs. 36%).

The conclusions of these findings are the measures for modernization of the course: Compared to UCBL, there is a lack of time in OMSU (on the programs that are involved in the MetaMath project) to study all algebra and geometry. So it is important to define learning goals and problems of the course more precisely and to follow them more strictly. It is necessary to collect study material according to the learning goals and problems defined. Study material should be even more illustrative than today and closer to IT-specificity. One of the ways to do this is to increase the number of computer programming labs; another is to use some software packages like GeoGebra. Students' preparatory work should be organized in a more effective way. The use of learning management systems (LMS) like Moodle or Math-Bridge will be very helpful here. The advantage of Math-Bridge is that this LMS is specially oriented to support mathematical courses.

Table 7.42 Outlines of discrete mathematics (DM) and algorithm mathematics (AM) courses at OMSU and TUT

Course information	OMSU	TUT
Bachelor/master level	Bachelor	Bachelor
Preferred year	1 or 2	2
Selective/mandatory	Mandatory	Mandatory
Number of credits	6 (216 h)	4 (105 h)
Teaching hours	108	49
Preparatory hours	108	65
Teaching assistants	No	Yes
Computer labs	Several programming tasks as homework	Available
Average number of students on the course	50	150
Average pass %	85–90%	90%
% of international students	Less than 10%	Less than 5%

7.4.3   Comparative Analysis of "Discrete Mathematics"

Discrete mathematics (DM) for the study programs ICS and SE is a fundamental mathematical course, so it is more theoretical than applied. But it is obvious that among all the theoretical courses it is the most applied and closest to the future of IT-professions of the students. OMSU tries to make the course more applied by giving the programming tasks to the students (for example, students must write a computer program which returns a breadth-first search in a connected graph). There are about 25 first year students in the SE program and 25 second year students in the ICS program, and all of them must study this course.

The corresponding course with our European partner Tampere University of Technology (TUT) is "Algorithm Mathematics" (AM). Course outlines can be seen in Table 7.42.

Prerequisite courses for Discrete Mathematics are secondary school mathematics and Algebra and Geometry. Follow-up courses are Mathematical Logic and Algorithm Theory, Theory of Automata and Formal Languages, Probability Theory and Statistics. The course of Discrete Mathematics is included in the group of mandatory mathematical courses that must be studied by all students of ICS and IT programs during the first years of study.

The Department of Applied Mathematics, Differential Equations and Theoretical Mechanics is responsible for this course for both programs, SE and ICS. Four full professors, 14 associate professors and 3 teachers work in this department.

The overall number of credits is 6. As mentioned before, in Russia we have 36 h in 1 credit, so the total amount is 216 h for this course. Among them are 36 h of lectures, 72 h of tutorials and 108 h of homework. Finnish credits contain less hours.

As for Algebra and Geometry, there is one 2-h lecture on Discrete Mathematics and two 2-h tutorials every week. During tutorials students solve some problems (fulfill computational tasks) under a teacher's direction. Students are given home tasks which must be done during preparatory hours. Computers are used also in controlling and grading the students' programming homework.

Course statistics for Discrete Mathematics is similar to that of Algebra and Geometry. There are about 25 ICS students and about 25 SE students attending the course; for these programs, lectures and tutorials on Discrete Mathematics are set separately. As said before, the number of students who do not finish the course is hard to give; after failing for the first time the student has two extra tries to pass the exam. Usually only Russian students study on ICS and SE programs. The average age of students attending the course is 19 (because for ICS students the course is in the second year of their study), and about 75% of the students are male. In OMSU there is no mandatory formal procedure of course rating; but as a rule students of IT-programs like the course because it is very "algorithmic" and close to programming.

The course of Discrete Mathematics is established for the first year or for second year students and is more theoretical than applied. The pedagogical methods used for this course are the same as for the course of Algebra and Geometry—and they are traditional. Students attend to lectures, fulfill tasks during tutorials and do their homework. Project work is not used widely, though some students interested in the course may be given additional large tasks (doing these tasks influences their rating). Similar to Algebra and Geometry, we do our best to make the learning process more interesting. For example, from time to time a group of students in the tutorial is divided into several subgroups and every subgroup fulfills some task. Searching a path in a connected graph by various methods is an example of such work. As mentioned before, this kind of work in subgroups is very competitive: students in different subgroups try to fulfill their task more quickly and more correctly than other subgroups. Blended learning is used episodically: some teachers use Moodle for distance learning. But for the students that have resident instruction (and here we discuss these students) it is more the exception than the rule.

Assessment and grading is similar to the course on "Algebra and Geometry", described above, and it is not repeated here.

7.4.3.1 Contents of the Course

The comparison is based on the SEFI framework [1]. Prerequisite competencies are presented in Table 7.43. Outcome competencies are given in Tables 7.44 and 7.45, 7.46, and 7.47.There are actually no other prerequisite competencies for Discrete Mathematics in OMSU. The course is taught from the very beginning of the field.

Table 7.43 Prerequisite competencies of discrete mathematics and algorithm mathematics courses at OMSU and TUT

Core 1		
Competency	OMSU	TUT
Proof	X	X

Table 7.44 Core 0 level outcome competencies of discrete mathematics and algorithm mathematics courses at OMSU and TUT

Core 0		
Competency	OMSU	TUT
Sets	X	X

Table 7.45 Core 1 level outcome competencies of discrete mathematics and algorithm mathematics courses at OMSU and TUT

Core 1		
Competency	OMSU	TUT
Sets	With some exceptions[a]	X
Mathematical induction and recursion	X	X
Graphs	X	
Combinatorics	X	

[a]Compare the algebra of switching circuits to that of set algebra and logical connectives; analyze simple logic circuits comprising AND, OR, NAND, NOR and EXCLUSIVE OR gates

Table 7.46 Core 2 level outcome competencies of discrete mathematics and algorithm mathematics courses at OMSU and TUT

Core 2		
Competency	OMSU	TUT
Number systems	Only: use Euclid's algorithm for finding the greatest common divisor.	
Algebraic operations	X	X
Recursion and difference equations	Only: define a sequence by a recursive formula.	X
Relations	X	X
Graphs	X	
Algorithms	Excluding: competencies due to NP and NP-complete problems.	X
Geometric spaces and transformations	Only: understand the group representation of geometric transformations.	

7.4.3.2 Summary of the Results

The main findings that are consequent from the comparison made are as follows: According to the amount of learning hours and to the list of topics covered the OMSU course of Discrete Mathematics is more extensive than the course in TUT.

Table 7.47 Core 3 level outcome competencies of discrete mathematics and algorithm mathematics courses at OMSU and TUT

Core 3		
Competency	OMSU	TUT
Combinatorics	Understanding the link between n-ary relations and relational databases. Ability to normalize database and to convert from 1NF to 2NF.	
Graph theory	Write a computer program that finds the components of connectivity, minimal spanning tree and so on.	
Algebraic structures	Using Shannon–Fano's and Huffman's methods to obtain optimal code; the LZW zipping algorithm, the Diffie-Hellman key exchange method; finding the RSA algorithm.	

Many topics that are covered in OMSU are not covered in TUT. Teachers pay attention to algebra within the course of Discrete Mathematics both in OMSU and in TUT. So it is a good practice for IT-students to learn about algebraic structures in the framework of Discrete Mathematics. Finnish colleagues widely use e-learning (Moodle), which helps to organize students' preparatory work more effectively. This experience is very useful. Another useful experience is that Moodle helps teachers to collect feedback from students after finishing the course.

These findings motivate the following modernization measures: In general, OMSU course of Discrete Mathematics is rather good and meets the requirements for teaching IT-students. But it will be useful to make it some more illustrative. It will be a good practice to continue using computer programming home works. In OMSU we should use e-learning to organize preparatory work for students in a more effective way. Math-Bridge and Moodle will help here. Also it will be suitable to collect some feedback from students.

7.5 Analysis of Mathematical Courses in TSU

Ilia Soldatenko and Alexander Yazenin
Tver State University (TSU), Information Technologies Department, Applied Mathematics and Cybernetics Faculty, Tver, Russia
e-mail: soldis@tversu.ru; Yazenin.AV@tversu.ru

Irina Zakharova
Tver State University (TSU), Mathematical Statistics and System Analysis Department, Applied Mathematics and Cybernetics Faculty, Tver, Russia
e-mail: zakhar_iv@mail.ru

Dmitriy Nikolaev
Tver State University (TSU), International Relations Center, Tver, Russia
e-mail: Nikolaev.DS@tversu.ru

7.5.1 Tver State University (TSU)

Tver State University is one of the largest scientific and educational centers in Central Russia. Responding actively to modern-day challenges, the institution of higher education is developing dynamically, while preserving tradition. TSU ensures the preparation of qualified specialists in the sphere of physico-mathematical, natural, human and social sciences, as well as of education and pedagogy, economy and administration among other areas. Tver State University is a classical institution with a total quantity of students equal to about 10 thousand, about half of which pursue STEM courses.

The Tver State University has had a long and difficult developmental path. The university's history starts on December 1, 1870, when, in Tver, a private pedagogical school named after P.P. Maximovich was opened. It was later on reformed in 1917 to become the Tver Teachers' Institute, after which it became the Kalinin Pedagogical Institute. Before the 1970s, tens of thousands of specialists graduated at the Pedagogical Institute with university qualifications. On September 1, 1971, an outstanding event took place in the Institute's history; it was renamed Kalinin State University.

In 1990, the Kalinin State University was renamed Tver State University. Its graduates work successfully at schools, scientific institutions, as well as in economic and social organizations. The university's scholars have also made a considerable contribution to making up and developing many scientific fields and research areas. Today, our personnel consists of about 600 professors, including 100 doctors, full professors and about 400 professors holding a PhD degree, as well as associate professors.

TSU is comprised of 12 faculties and 2 institutes, which are the following:

- Institute of Pedagogical Education and Social Technologies,
- Institute of Economics and Management,
- Faculty of Biology,
- Faculty of History,
- Faculty of Mathematics,
- Faculty of Geography and Geo-ecology,
- Faculty of Foreign Languages and International Communication,
- Faculty of Applied Mathematics and Cybernetics,
- Faculty of Psychology,
- Faculty of Sport,
- Physico-Technical Faculty,
- Faculty of Philology,
- Faculty of Chemistry and Technology,
- Faculty of Law.

Our personnel consists of about 600 professors, including 100 doctors, full professors and about 400 professors holding a PhD degree, as well as associate professors.

The main directions of research and development at the University are carried out in the field of natural and exact sciences: mathematics, mechanics, physics, chemistry, biology, geo-ecology and computer science. There is also a lot of research in the fields of the humanities and social sciences, such as sociology, linguistics, literature, history, economics, state and law, as well as in protection of the environment, human ecology and demography. At the University, there are over 20 scientific schools carrying out research in relevant scientific topics of natural sciences and humanities within 15 fields of study. Their activities are recognized internationally, as well as domestically.

TSU maintains close ties with more than 30 universities in Europe, the USA and the Commonwealth of Independent States, and it carries out exchange programs, and it provides education to international students and actively participates in various international educational programs. Tver State University's long-standing partners include the University of Osnabruck and the University of Freiburg (Germany), the University of Montpellier and the University of Clermont-Ferrand (France), the University of Turku and the University of Joensuu (Finland), the University of Ghent (Belgium), the University of Xiamen (China), St. Cyril and St. Methodius University of Veliko Tarnovo (Bulgaria), the University of Glasgow (UK), etc. Thanks to many years of international cooperation, Tver State University has established educational and cultural ties with these institutions and has increased exchanges in the field of scientific research. TSU is one of the few higher education institutions that develops international academic mobility.

Annually, about 100 students from TSU's different departments attend a course of study for one semester at universities in the Federal Republic of Germany, France, Finland, Bulgaria, Poland, and the USA. Students attend classes according to the profile of their learning, and they take exams. Students perfect their knowledge of foreign languages, acquire the invaluable experience of studying at a higher learning institution abroad, and make new and interesting friends.

The number of TSU students studying for a semester at the University of Osnabruck at their own expense keeps increasing each year. The university's involvement in the "East–West" research faculty exchange program jointly financed by DAAD and the University of Osnabruck has facilitated the creation of close-knit research teams in the field of mathematics, geography, chemistry, botany and the publishing of books, articles and other publications.

Since 2000, students coming from different universities in Finland have taken part in inclusive semester courses at the Department of Russian as a Foreign Language. The project is financed by the Ministry of Education of Finland. Since 2005 TSU has been a participant of "FIRST" program (Russia–Finland Student Exchange Program). A similar program has been conducted with UK universities and the cooperation of the RLUS Company.

The process of TSU integration into the global educational space is also realized through the involvement in different international educational schemes. Each year, more than 200 students and post-graduate students from foreign universities (including the CIS and the Baltic nations) take a course at Tver State University.

In recent years the university has significantly expanded its cooperation with the Oxford–Russia Fund. This project is supervised by the TSU Inter-University Centre for International Cooperation. Annually 120 TSU students are awarded scholarships by the Fund. The University was also given access to the electronic library of the University of Oxford. TSU was among the first 10 partners of the Foundation to gain access to online versions of 500 British titles. The project also envisages the TSU Library receiving books on art, languages, history, etc.

The cooperation with the Fulbright Foundation allows TSU to annually host US guest speakers delivering lectures on international affairs, global terrorism, etc. TSU faculty and students are actively involved in different educational and research programs and projects financed by the European Union, the Ford Foundation, CIMO, the DAAD, the IREX, etc.

7.5.2   Mathematics Education in Tver State University

The Faculty of Applied Mathematics and Cybernetics (AM&C) and the Mathematical Faculty are responsible for conducting mathematical courses at the university. The Mathematical Faculty focuses on pedagogical programs; AM&C on applied mathematics. The Faculty of Applied Mathematics and Cybernetics was founded in 1977, though the specialization "Applied Mathematics" was opened in 1974. It is in many areas the leading educational department in the University. It includes more than 32 full-time teachers, including 19 candidates and 10 doctors of sciences who have major scientific achievements in their respective fields of expertise. Teaching staff also consist of representatives of employers, who have extensive practical experience.

Currently the Faculty has four educational programs:

- Applied Mathematics and Computer Science,
- Fundamental Computer Science and Information Technologies,
- Computer Science in Business,
- Applied Computer Science.

In addition to these programs, the Faculty provides training in mathematics and appropriate applied disciplines in other faculties. The Faculty has four departments:

1. Information Technologies department (fields of expertise: intellectual information systems, fuzzy systems and soft computing technologies; theory of possibilities; probabilistic and probabilistic optimization and decision-making; portfolio theory under conditions of hybrid uncertainty; processing and recognition of signals and images; multimedia technologies).
2. Computer Science department (fields of expertise: theoretical programming, theory of finite models, theory of multi-agent systems, theoretical linguistics, the development of expert systems; databases).

3. Mathematical Statistics and System Analysis department (fields of expertise: theory of sustainable and natural exponential probability distributions of data; application of probabilistic and statistical methods in econometrics, financial and actuarial mathematics, analysis of telecommunication networks; choice theory, multicriteria decision making under uncertainty).
4. Mathematical Modeling and Computational Mathematics department (fields of expertise: theoretical basis of mathematical modeling of complex systems; development of mathematical models of critical states of nonlinear dynamical systems, assessment of the safety performance of these systems; mathematical models and methods for the identification of objects with large interference and decision making under uncertainty, analysis and solution finding using mathematical modeling applications: solution of problems of nonlinear elastic and viscoelastic materials; solving problems of geophysical hydrodynamics and hydrothermodynamics; solving problems of heat conduction; analysis of the structures of insurance funds, development of recommendations on the organization of insurance funds; development and application of optimization methods for solving economic problems; development and implementation of a measurement system of numerical methods for optimal digital video and audio signals).

The faculty also has three scientific schools: Fuzzy systems and soft computing, Mathematical modeling, Theoretical foundations of computer science.

7.5.3   Comparative Analysis of "Probability Theory and Mathematical Statistics"

"Probability Theory and Mathematical Statistics" is a mandatory course at Tver State University; it combines theoretical and applied approaches. Mathematics plays a key role in the course, and it in its turn forms a foundation for several other applied special disciplines in corresponding educational programs. The course was compared with two courses at Tampere University of Technology (TUT): "Probability Calculus" and "Statistics". The course outlines can be seen in Table 7.48.

Prerequisite courses at TSU are Linear Algebra, Calculus and Differential Equations, and follow-up courses are Possibility Theory and Fuzzy Logic, Econometrics, Theory of Stochastic Processes and Methods of Socio-Economic Forecasting. At TUT, prerequisite courses are Engineering Mathematics 1–4, and follow-up courses are all the other courses provided by the department. In Tver State University this course is included in the group of core mandatory mathematical courses of the corresponding study programs and is taught by the Applied Mathematics and Cybernetics department. In TUT the course is taught by Department of Mathematics and it is also included in the group of mandatory mathematical courses.

The teaching of the course at TSU is more theory-based and unfortunately does not include any innovative pedagogical methods and tools, such as blended learning,

Table 7.48 Outlines of the courses on probability theory and mathematical statistics at TSU and TUT

Course information	TSU	TUT
Bachelor/master level	Bachelor	Bachelor
Preferred year	2–3	2
Selective/mandatory	Mandatory	Both
Number of credits	10	4 + 4
Teaching hours	148	42 + 42
Preparatory hours	220	132
Teaching assistants	0–1	1–4
Computer labs	Practice with MATLAB, Excel, C++	Practice in R-software
Average number of students on the course	85	200
Average pass %	90%	90%
% of international students	10%	< 5%

flipped classroom, MOOCs, project-based learning, inquiry-based learning, collaborative learning, gamification. In TUT on the other hand one can find blended learning, collaborative learning, project-based approach and active use of modern TEL tools for administration, teaching and assessment purposes. Not all modern pedagogical technologies are used in TUT but those that do exist in educational process are applied widely and successfully.

Assessment, testing and grade computation do not differ from what is described in Chap. 3. At TSU, as in all Russian universities, a general rating system is implemented. Rating is the sum of points for all courses taken during the whole of 4 years of education. The rating comes into play when it is time for a student to choose his/her major (graduating chair), which has influences on the further curriculum (set of special professional courses) and Bachelor thesis topics. The exam for the course is verbal; during the semester lecturer performs several written tests. A student can receive 100 points for the discipline, 60 of which come from activities during semester and 40 from exam work. The semester is also divided into two modules, each of which gives a maximum of 30 points. The grading systems is the following: less than 50 points is unsatisfactory (grade "2"), from 50 to 69 points satisfactory (grade "3"), from 70 to 84 good (grade "4"), more than 84 excellent (grade "5").

TUT on the other hand does not have a general rating system and uses six grades according to European ECTS scale to assess results of education in each course: excellent (grade A) 5 points; very good (grade B) 4 points; good (grade C) 3 points; very satisfactory (grade D) 2 points; satisfactory (grade E) 1 point; unsatisfactory (grade F) 0 points.

As regards educational software and TEL systems in TSU in the course: MATLAB and Excel are used in tutorials. Lecturers teach students how to use the tools, but these are not mandatory for solving practical and assessment tasks. A student may also choose to write a program in any high-level multi-purpose language, which he or she knows already. In TUT, educational software and TEL systems are used in the following ways: Moodle for file sharing and course information, POP for course and exam enrollment and course grades, and R is used as support for exercises. Furthermore, students in TUT form their individual learning paths from the very beginning of their studies with the help of specialized software. There is an electronic catalog of courses from which a student should choose an appropriate number of courses with the needed amount of credits. This is significantly different from what we have in Tver State University as well as in all other Russian universities.

7.5.3.1 Contents of the Course

The comparison is based on the SEFI framework [1]. Prerequisite competencies are presented in Table 7.49. Outcome competencies are given in Tables 7.50, 7.51, and 7.52.

7.5.3.2 Summary of the Results

Comparative analysis of the disciplines shows that thematic contents and learning outcomes are almost identical. The difference is observed in the number of hours. One should also note the active use of information technologies and, in particular, e-learning systems in TUT. This allows this university to take out some of the material

Table 7.49 Prerequisite competencies of the probability theory and mathematical statistics courses at TSU and TUT

Core 1		
Competency	TSU	TUT
Data handling	Excluding: calculate the mode, median and mean for a set of data items	
Arithmetic of real numbers	X	X
Algebraic expressions and formulas	X	X
Functions and their inverse	X	X
Sequences, series, binomial expansions	X	X
Logarithmic and exponential functions	X	X
Indefinite integration	X	X
Definite integration, applications to areas and volumes	With some exceptions[a]	With some exceptions[a]
Sets	X	X

[a]Use trapezoidal and Simpson's rule for approximating the value of a definite integral

Table 7.50 Core 0 level outcome competencies of the probability theory and mathematical statistics courses at TSU and TUT

Core 0		
Competency	TSU	TUT
Calculate the mode, median and mean for a set of data items	X	X
Define the terms 'outcome', 'event' and 'probability'	X	X
Calculate the probability of an event by counting outcomes	X	X
Calculate the probability of the complement of an event	X	X
Calculate the probability of the union of two mutually exclusive events	X	X
Calculate the probability of the union of two events	X	X
Calculate the probability of the intersection of two independent events	X	X

Table 7.51 Core 1 level outcome competencies of the probability theory and mathematical statistics courses at TSU and TUT

Core 1		
Competency	TSU	TUT
Calculate the range, inter-quartile range, variance and standard deviation for a set of data items	X	X
Distinguish between a population and a sample	X	X
Know the difference between the characteristic values (moments) of a population and of a sample	X	X
Construct a suitable frequency distribution from a data set	X	
Calculate relative frequencies	X	
Calculate measures of average and dispersion for a grouped set of data	X	
Use the multiplication principle for combinations	X	X
Interpret probability as a degree of belief	X	
Understand the distinction between a priori and a posteriori probabilities	X	X
Use a tree diagram to calculate probabilities	X	
Know what conditional probability is and be able to use it (Bayes' theorem)	X	X
Calculate probabilities for series and parallel connections	X	
Define a random variable and a discrete probability distribution	X	X
State the criteria for a binomial model and define its parameters	X	X
Calculate probabilities for a binomial model	X	X
State the criteria for a Poisson model and define its parameters	X	X
Calculate probabilities for a Poisson model	X	X
State the expected value and variance for each of these models	X	X
Understand that a random variable is continuous	X	X
Explain the way in which probability calculations are carried out in the continuous case	X	
Relate the general normal distribution to the standardized normal distribution	X	X
Define a random sample	X	X
Know what a sampling distribution is	X	X
Understand the term 'mean squared error' of an estimate	X	
Understand the term 'unbiasedness' of an estimate	X	

Table 7.52 Core 2 level outcome competencies of the probability theory and mathematical statistics courses at TSU and TUT

Core 2		
Competency	TSU	TUT
Compare empirical and theoretical distributions	X	X
Apply the exponential distribution to simple problems	X	
Apply the normal distribution to simple problems	X	X
Apply the gamma distribution to simple problems	X	X
Understand the concept of a joint distribution	X	X
Understand the terms 'joint density function', 'marginal distribution functions'	X	X
Define independence of two random variables	X	X
Solve problems involving linear combinations of random variables	X	X
Determine the covariance of two random variables	X	X
Determine the correlation of two random variables	X	X
Realize what the normal distribution is not reliable when used with small samples	X	
Use tables of the t-distribution	X	X
Use tables of the F-distribution	X	X
Use the method of pairing where appropriate	X	X
Use tables for chi-squared distributions	X	X
Decide on the number of degrees of freedom appropriate to a particular problem	X	X
Use the chi-square distribution in tests of independence (contingency tables)	X	X
Use the chi-square distribution in tests of goodness of fit	X	
Set up the information for a one-way analysis of variance	X	X

for independent study and focus on really difficult topics of the discipline. E-learning systems also allow one to automate and, as a result, simplify the knowledge assessment process. This automation seems to be important for TUT, whose class sizes substantially exceed the size of study groups in the TSU.

All this suggests the need for more active use of e-learning systems, as well as blended learning methodology in the educational process in Tver State University. It is also worth noting that in TUT part of the basic (input) material is moved to a bridging course "Mathematics Basic Skills Test & Remedial Instruction". The material in this course is designed primarily for successful mastering of the engineering mathematics courses. Nevertheless, this experience should also be useful for Tver State University. In particular, a bridging course "Basics of Elementary Mathematics" should be created, which will include the material from the following topics of mathematics: Set theory, elementary functions and their graphs, series and their properties, elements of combinatorics, equations and inequalities.

Table 7.53 Outline of possibility theory and fuzzy logic course at TSU

Course information	TSU
Bachelor/master level	Bachelor
Preferred year	3
Selective/mandatory	Selective
Number of credits	4
Teaching hours	68
Preparatory hours	76
Teaching assistants	1–2
Computer labs	Not used
Average number of students on the course	50
Average pass %	95%
% of international students	17%

7.5.4   Comparative Analysis of "Possibility Theory and Fuzzy Logic"

Because of the fact that the discipline "Possibility Theory and Fuzzy Logic" is not widely spread among both domestic and foreign universities and is a particular feature of Tver State University, the comparative analysis was performed with the course "Probability Theory and Mathematical Statistics". However, because of the relative proximity of the disciplines all its conclusions including recommendations on the modernization of the course are applicable to "Possibility Theory and Fuzzy Logic", as well. Below is the profile of the second course in terms of its structure and prerequisite SEFI competencies. It is an elective course which combines both theoretical and applied approaches. Mathematics plays a key role in it. The course outline can be seen in Table 7.53.

Prerequisite courses are Probability Theory and Methods of Optimisation and Decision Making. There are no follow-up courses for "Possibility Theory and Fuzzy Logic", because this course is included in the elective part of the professional cycle of the corresponding educational programs. It is taught by lecturers from the Applied Mathematics and Cybernetics department. The teaching of this course is more theory-based and classical without wide support of e-learning tools and methods.

7.5.4.1   Contents of the Course on Possibility Theory and Fuzzy Logic

The comparison is based on the SEFI framework [1]. Prerequisite competencies are presented in Tables 7.54, 7.55, and 7.56.

Table 7.54 Core 0 level prerequisite competencies of the course on possibility theory and fuzzy logic at TSU

Core 0	
Competency	TSU
Arithmetic of real numbers	X
Algebraic expressions and formulas	X
Linear laws	X
Functions and their inverses	X
Logarithmic and exponential functions	X
Indefinite integration	X
Proof	X
Sets	X
Coordinate geometry	With some exceptions[a]
Probability	X

[a]Find the angle between two straight lines, recognize and interpret the equation of a circle in standard form and state its radius and center, convert the general equation of a circle to standard form, derive the main properties of a circle, including the equation of the tangent at a point, recognize the parametric equations of a circle, use polar coordinates and convert to and from Cartesian coordinates

Table 7.55 Core 1 level prerequisite competencies of the course on possibility theory and fuzzy logic at TSU

Core 1	
Competency	TSU
Rational functions	With some exceptions[a]
Functions	X
Solution of simultaneous linear equations	X
Simple probability	X
Probability models	X

[a]Obtain the first partial derivatives of simple functions of several variables, use appropriate software to produce 3D plots and/or contour maps

Table 7.56 Core 2 level prerequisite competencies of the course on possibility theory and fuzzy logic at TSU

Core 2	
Competency	TSU
Linear optimization	With some exceptions[a]

[a]Understand the meaning and use of slack variables in reformulating a problem, understand the concept of duality and be able to formulate the dual to a given problem

Unfortunately SEFI Framework does not have learning outcomes suitable for "Possibility Theory and Fuzzy Logic". After successful completion of the course, a student should have to master:

- the mathematical apparatus of the possibility theory and knowledge representation in computer science,

- skills to model uncertainty of probabilistic type in decision-making problems,
- methods of knowledge representation with elements of possibilistic uncertainty,

they should be able to:

- use this mathematical apparatus in the development of fuzzy decision support systems,
- apply mathematical apparatus of possibility theory in modern information technologies,
- use soft computing technologies for solving applied problems,

they should know:

- elements of fuzzy sets theory, fuzzy logic theory and modern possibility theory, soft computing technologies,
- fundamental concepts and system methodologies in the field of information technologies based on soft computing,
- principles of construction of fuzzy decision support systems.

7.5.4.2 Summary of the Results

Because of the relative proximity of this discipline to Probability Theory all the conclusions for the latter discipline are applicable to "Possibility Theory and Fuzzy Logic" as well.

Reference

1. SEFI (2013), "A Framework for Mathematics Curricula in Engineering Education". (Eds.) Alpers, B., (Assoc. Eds) Demlova M., Fant C-H., Gustafsson T., Lawson D., Mustoe L., Olsson-Lehtonen B., Robinson C., Velichova D. (http://www.sefi.be).

Chapter 8
Case Studies of Math Education for STEM in Georgia

8.1 Analysis of Mathematical Courses in ATSU

Tea Kordzadze
Akaki Tsereteli State University (ATSU), Department of Mathematics, Kutaisi, Georgia

Tamar Moseshvili
Akaki Tsereteli State University (ATSU), Department of Design and Technology, Kutaisi, Georgia

8.1.1 Kutaisi Akaki Tsereteli State University (ATSU)

The history of Akaki Tsereteli State University (ATSU) started eight decades ago and now it is distinguished with its traditions throughout Georgia and holds an honorable place in the business of cultural, intellectual and moral education of the Georgian nation. According to the Georgian government's resolution #39, February 23, 2006 the legal entities of public law Kutaisi Akaki Tsereteli State University and Kutaisi N. Muskhelishvili State Technical University were combined, the educational status being determined to be the university and the new entity was named Akaki Tsereteli State University. ATSU was merged with Sukhumi Subtropical Teaching University in 2010.

ATSU became one of the largest universities, with a wide spectrum of academic (on BA, MA, and PhD levels), professional teaching programs and research

© The Author(s) 2018
S. Pohjolainen et al. (eds.), *Modern Mathematics Education for Engineering Curricula in Europe*, https://doi.org/10.1007/978-3-319-71416-5_8

fields. Today in ATSU there are about 11,000 students, 9 faculties and 11 STEM disciplines.

8.1.2   Comparative Analysis of "Calculus 1"

Calculus 1 is a mandatory course for students of STEM specializations. This course was developed for engineering students. It is theoretical course with practical examples connected to real life problems. It was compared with "Engineering Mathematics 1" (EM1), which is similar to a course at Tampere University of Technology. The outlines of the courses are presented in Table 8.1. The Department of Mathematics is responsible for the course at Akaki Tsereteli State University. There are 4 full-time professors, 18 full-time associate professors, and 5 teachers.

There are no prerequisite courses. The course size is 5 credits, and it requires on average 125 h of work (25 h per credit). The credits are divided among different activities as follows: lectures 30 h, tutorials 30 h, homework 58 h, consultation 3 h, and exam 4 h. On the course organizer side, there are 60 contact teaching hours. From one to two teaching assistants work on the course to grade exams and teach tutorials. Unfortunately, there are no computer labs available for use on the course.

There are approximately 200 students on the course. The average number of students that finish the course is 170 (85%). The amount of international students is less than 1%.

The pedagogical comparison of the course shows that the teaching is very much theory based. Professors of the mathematics department are delivering the course, and they do not use modern teaching methods such as blended learning, flipped classroom, project or inquiry based learning etc.

Table 8.1 Outlines of Calculus 1 and EM1 course at ATSU and TUT

Course information	ATSU	TUT
Bachelor/Master level	Bachelor	Bachelor
Preferred year	1	1
Elective/mandatory	Mandatory	Mandatory
Number of credits	5	5
Teaching hours	60	56
Preparatory hours	58	75
Teaching assistants	1–2	1–3
Computer labs	Available	Available
Average number of students on the course	200	200
Average pass%	85%	90%
% of international students	Less than 1%	Less than 5%

The course's maximum evaluation equals 100 points. A student's final grade is obtained as a result of summing the midterm evaluation earned per semester and final exam evaluation results.

Assessment Criteria Students are evaluated in a 100-point system in which 45 points are given from midterm assessments, 15 points from Activities and 40 points from the final exam. Midterm assessments include the following components: Work in group (40 points) and two midterm examinations (20 points). Within the frames of 40-point assessment for working in a group students can be given an abstract/review work with 10 points. Students are obliged to accumulate no less than 11 points in midterm assessments and no less than 15 points at the exams. The course will be considered covered if students receive one of the following positive grades: (A) Excellent: 91 points and more; (B) Very good: 81–90 points; (C) Good: 71–80 points; (D) Satisfactory: 61–70 points; (E) Sufficient: 51–60 points. (FX) No pass— in the case of getting 41–50 points students are given the right to take the exam once again. (F) Fail—with 40 points or less, students have to do the same course again.

8.1.2.1 Contents of the Course

The comparison is based on SEFI framework [1]. Prerequisite competencies are presented in Table 8.2. Outcome competencies are given in Table 8.3.

Table 8.2 Core 0-level prerequisite competencies for Calculus 1 (ATSU) and EM1 (TUT) courses

Core 0		
Competency	ATSU	TUT
Arithmetic of real numbers	X	X
Algebraic expressions and formulas	X	X
Linear laws	X	X
Quadratics, cubics, polynomials	X	X
Functions and their properties	X	X

Table 8.3 Core 1-level outcome competencies for Calculus 1 (ATSU) and EM1 (TUT) courses

Core 1		
Competency	ATSU	TUT
Sets, operations on sets	X	X
Functions and their properties	X	X
Logarithmic and trigonometric functions	X	X
Limits and continuity of a function	X	X
Derivative	X	X
Definite integral and integration methods	X	
Minima and maxima	X	X
Indefinite integrals	X	
Definite integrals	X	

8.1.2.2   Summary of the Results

The main differences between Finnish (TUT) and Georgian (ATSU) courses that in TUT teaching is more intense. ATSU has 30 lectures and 30 tutorials—but TUT does all of this in 7 weeks, whereas ATSU uses 15 weeks. On the other hand Calculus 1 covers more mathematical areas than TUT's EM1. TUT also uses plenty of TEL and ICT technologies to support the teaching, ICT is not used in the study process in ATSU. Finnish students also answer in a pen and paper examination as ATSU student, but ATSU students must pass three exams; two midterm exams and one final exam, with no pure theoretical questions. Due to this comparison ATSU changed syllabus in Calculus 1. To achieve SEFI competencies ATSU modernized the syllabus by integration of Math-Bridge and GeoGebra tools in the study process. Modernization of Syllabus was done on September 2016. In 2016 ATSU has done pre- and post-testing in Calculus 1. During post-testing Math-Bridge tools were used for testing and analyzing results.

8.1.3   Comparative Analysis of "Modeling and Optimization of Technological Processes"

The course "Modeling and Optimization of Technological Processes" is offered for Master students of technological engineering faculty in the following engineering programs: Technology of Textile Industry, Food Technology, Technology of Medicinal Drugs, Engineering of Environmental Pollution and Ecology of Wildlife Management. ECTS credits for the course are 5 (125 h). The teaching language is Georgian, the average number of students is 30. It was compared with a corresponding course "Mathematical Modelling" from TUT. The course outlines are seen in Table 8.4.

Specific engineering departments are responsible for the course. A professor in each department teaches this course.

Table 8.4 Outlines of modeling courses at ATSU and TUT

Course information	ATSU	TUT
Bachelor/Master level	Master	Both
Preferred year	1	1
Elective/mandatory	Mandatory	Elective
Number of credits	5	5
Teaching hours	48	30
Preparatory hours	77	108
Teaching assistants	1	1
Computer labs	Yes	Yes
Average number of students on the course	30	60
Average pass%	75%	95%
% of international students	0%	Less than 5%

Prerequisite courses are the Bachelor courses of Higher Mathematics. The course size is 5 credits, and it requires on average 125 h of work (25 h for each credit). The credits are divided among different activities as follows: lectures 15 h, tutorials 30 h, independent work 77 h, exam 3 h.

On the course organizer side, there are 48 contact teaching hours. There is one teaching assistant to work on the course to grade exams and teach tutorials. There is one computer lab available for use on the course. There are approximately 30 students on the course. The average number of students that finish the course is 20 (75%). There are no international students in the technological engineering faculty. The pedagogical comparison of the course shows that the teaching process is not modern. Professors delivering the course do not use modern teaching methods such as blended learning, flipped classroom, project or inquiry based learning etc.

Grading is done on a 100 point scale with 51 points being the passing level. A score between 41 and 50 allows the student a new attempt at the exam, and every 10 point interval offers a better grade with 91–100 being the best.

Pen and paper exams are conducted three times in a semester.

- First midterm exam comprises 1–5 weeks materials and is conducted after the 5th week in compliance with Grading Center schedule.
- Second midterm exam comprises 7–11 weeks materials and is conducted after the 11th week in compliance with Grading Center schedule.
- Final exam is conducted after the 17–18th week.

For the final evaluation the scores of the midterm tests and independent work are summed up.

8.1.3.1   Contents of the Course

The comparison is based on SEFI framework [1]. Outcome competencies are given in Table 8.5.

8.1.3.2   Summary of the Results

The main differences between Finnish (TUT) and Georgian (GTU) courses are the following: The amount of contact hours in ATSU are 48 h. This consists of lectures,

Table 8.5 Core 2-level outcome competencies of the modeling courses at ATSU and TUT

Core 2		
Competency	ATSU	TUT
Simple linear regression	X	X
Multiple linear regression and design of experiments	X	X
Linear optimization	X	X
The simplex method	X	
Nonlinear optimization	X	

15 h (1 h per week), and practical work, 30 h, and midterm and final exam, 3 h. The amount of independent work is 77 h (62%).

In TUT the course is given as a web based course. Two hours of video lectures are implemented per week. 108 h are for laboratory work/tutorials. All of it is group work for weekly exercises and the final project. 100% of the student's time is for homework, which is mandatory for the students.

There are big differences of teaching methods between our courses. In ATSU the lectures are implemented by using verbal or oral methods: giving the lecture materials to the students orally according to the methods of questioning and answering, interactive work, explaining theoretical theses on the bases of practical situation simulation.

In TUT the course is given as a web based course. Different universities around Finland participate in the project. Each week a different university is responsible for that week's topic. The main coordination is done by TUT. Students form groups in each university and work on the given tasks as teams.

Modern lecture technology is used in TUT: e-Learning, with a hint of blended learning, Moodle, MATLAB or similar software, online lecture videos. Moodle is used for file sharing, course information, peer assessment of tasks and MATLAB for solving the exercises.

Video lectures online, weekly exercises are done in groups, posted online and then reviewed and commented on by other groups. Students are awarded points for good answers and good comments. At the end of the course a final project is given to the students to undertake. The final project work is assessed by the other students and by the course staff, and it is presented in a video conference.

ATSU will prepare new syllabi for modernized courses that are more in line with European university courses. This will be done in order to better prepare the Master students of ATSU for their future careers. The Georgian educational system teaches less mathematics on the high school level, and thus ATSU has to design courses that upgrade students' knowledge in these topics as well.

As the general level of the students is quite low, ATSU considers that the increase of the credits in mathematics on the Bachelor level is necessary. Also, the use of different software packages to support learning (Math-Bridge, GeoGebra, etc.) will increase the quality of knowledge of our students.

8.2   Analysis of Mathematical Courses in BSU

Vladimer Baladze, Dali Makharadze, Anzor Beridze, Lela Turmanidze, and Ruslan Tsinaridze
Batumi Shota Rustaveli State University (BSU), Department of Mathematics, Batumi, Georgia
e-mail: dali_makharadze@mail.ru

8.2.1   *Batumi Shota Rustaveli State University (BSU)*

Batumi Shota Rustaveli State University is deservedly considered to be one of the most leading centers of education, science and culture in Georgia. Educational and scientific activities of BSU go back to 1935. From the date of its establishment it has been functioning as the pedagogical institute providing Western Georgia with pedagogically educated staff for 55 years.

In 1990 the classical university with the fundamental, humanitarian and social fields was established on the basis of the pedagogical institute. In 2006, based on the decision of the Georgian government, scientific-research institutes and higher education institutions of various profiles located in the territory of Autonomous Republic of Ajara, joined Shota Rustaveli State University.

Currently 6000 students study at vocational, BA, MA and PhD educational programs. The process of education and research is being implemented by 244 professors, 55 scientists and researchers and 387 invited specialists.

BSU offers the students a wide-range choice in the following programs at all three levels: 43 BA, 44 MA, 28 PhD and 2 one-level programs. Based on the labor market demand some faculties implement vocational programs as well.

Shota Rustaveli State University is located in Batumi and its outskirts area, having six campuses. The university comprises seven faculties and three scientific-research institutes:

- Faculties:
    Faculty of Humanities;
    Faculty of Education;
    Faculty of Social and Political Sciences;
    Faculty of Business and Economics;
    Faculty of Law;
    Faculty of Natural Sciences and Health Care;
    Faculty of Physics-Mathematics and Computer Sciences;
    Faculty of Technology;
    Faculty of Tourism.
- Scientific-research institutes:
    Niko Berdzenishvili Institute (Direction of Humanities and Social Studies);
    Institute of Agrarian and Membrane Technologies;
    Phytopathology and Biodiversity Institute.

8.2.2 Comparative Analysis of "Linear Algebra and Analytic Geometry (Engineering Mathematics I)"

"Linear Algebra and Analytic Geometry" (or "Engineering Mathematics I)" (EMI) is a theoretical course with approximately 50 students. The course is a first year course for engineering students at the Faculty of Technology and is a mandatory course of engineering programs in BSU. The course is compared with the "Engineering Mathematics 1" (EM1) course at Tampere University of Technology. The course outlines are presented in Table 8.6.

The Department of Mathematics is responsible for the course. The staff of this department consists of three full-time professors, six full-time associate professors, four full-time assistant professors and four teachers. The Department of Mathematics conducts the academic process within the frames of the educational programs of the faculty Technologies, of the faculty Physics, Mathematics and Computer Sciences as well as other faculties.

The course does not have prerequisite courses. The course size is 5 credits, and it requires on average 125 h of student's work (25 h for each credit). The credits are divided among different activities as follows: lectures 15 h, tutorials 30 h, homework 80 h. Students should use about 30 h to prepare for their exam.

On the course organizer side, there are 60 contact teaching hours. From one to three teaching assistants work on the course to grade exams and teach tutorials. Unfortunately, there are no computer labs available for use on the course.

There are 50 students on the course. The average number of students that finish the course is 41 (82%). The amount of international students is 0%.

The pedagogical comparison of the course shows that the teaching is very much theory based. Professors of mathematics department are delivering the course, and they do not use modern teaching methods such as blended learning, flipped classroom, project or inquiry based learning etc.

Table 8.6 Outlines of EM I (BSU) and EM1 (TUT) courses

Course information	BSU	TUT
Bachelor/Master level	Bachelor	Bachelor
Preferred year	1	1
Elective/mandatory	Mandatory	Mandatory
Number of credits	5	5
Teaching hours	45	57
Preparatory hours	80	76
Teaching assistants	1–3	1–3
Computer labs	No	Yes
Average number of students on the course	50	200
Average pass%	82%	90%
% of international students	No	Less than 5%

Assessment Criteria Students are evaluated in a 100-point system in which 60 points are given from midterm assessments and 40 points from the final exam. Midterm assessments include the following components: Work in group (40 points) and two midterm examinations (20 points). Within the frames of 40-point assessment for working in a group students can be given an abstract/review work with 10 points. Students are obliged to accumulate no less than 11 point in midterm assessments and no less than 21 points at the exams. The course will be considered covered if students receive one of the following positive grades:

- (A) Excellent: 91 points and more.
- (B) Very good: 81–90 points.
- (C) Good: 71–80 points.
- (D) Satisfactory: 61–70 points.
- (E) Sufficient: 51–60 points.

(FX) No pass—in the case of getting 41–50 points students are given the right to take the exam once again.

(F) Fail—with 40 points or less, students have to do the same course again.

"Linear Algebra and Analytical Geometry/Engineering Mathematics I" at BSU does not have any TEL tools available on the course. Therefore, no TEL tools are used to support learning.

8.2.2.1 Contents of the Course

The comparison is based on SEFI framework [1]. Prerequisite competencies are presented in Table 8.7. Outcome competencies are given in Tables 8.8 and 8.9.

Table 8.7 Core 0-level prerequisite competencies for EMI (BSU) and EM1 (TUT) courses

Core 0		
Competency	BSU	TUT
Arithmetic of real numbers	X	X
Algebraic expressions and formulas	X	X
Linear laws	X	X
Quadratics, cubics, polynomials	X	X
Functions and their inverses	X	X
Sequences, series, binomial expansions	Excl. series	X
Logarithmic and exponential functions	X	X
Geometry	X	X
Trigonometry	X	X
Trigonometric identities	X	X

Table 8.8 Core 0-level
outcome competencies for
EMI (BSU) and EM1 (TUT)
courses

Core 0		
Competency	BSU	TUT
Binomial expansions	X	X
Sets	X	X
Co-ordinate geometry	X	X

Table 8.9 Core 1-level
outcome competencies for
EMI (BSU) and EM1 (TUT)
courses

Core 1		
Competency	BSU	TUT
Sets	X	X
Complex number	X	X
Mathematical logic	X	X
Mathematical induction	X	X
Conic sections	X	X
3D co-ordinate geometry	X	
Vector arithmetic	X	X
Vector algebra and applications	X	
Matrices and determinants	X	
Solution of simultaneous linear equations	X	

8.2.2.2   Summary of the Results

The main differences between Finnish (TUT) and Georgian (BSU) courses are the
following: in TUT, teaching is more intense and covers less topics than BSU. The
overall hours are somewhat different; TUT has 35 h of lectures and 21 h of tutorials;
BSU has 15 lectures and 30 tutorials, and TUT does all of this in 7 weeks, whereas
BSU uses 15 weeks. TUT also uses plenty of TEL and ICT technologies to support
their teaching, BSU does not. Finally, the exams are somewhat different. Finnish
students answer in a pen and paper exam, BSU students must pass three exams with
theoretical questions, but with no proofs.

The main drawbacks of the old mathematics syllabi at BSU were that mostly the
theoretical mathematical aspects were treated and the corresponding examinations
contained only purely mathematical questions.

Moreover, it should be specially mentioned that in the BSU the mathematical
syllabus "Engineering Mathematics 1" (= "Linear Algebra and Analytical Geometry") in engineering BSc programs does not include the following topics: Elements
of Discrete Mathematics, Surfaces of second order. Therefore, modernization of the
syllabus of "Linear Algebra and Analytical Geometry" is very desirable.

BSU will prepare new syllabi for modernized courses that are more in line with
European technical university courses. This will be done in order to better prepare
the students of BSU for their future careers. However, modernizing courses will
not be trivial, since the university has to make up for the different levels of skills
of European and Georgian enrolling students. The Georgian educational system
teaches less mathematics on the high school level, and thus BSU has to design
courses that upgrade students' knowledge in these topics as well.

As the overall level of students is relatively low, BSU finds that implementing remedial mathematics courses is necessary. This could be done using the Math-Bridge software.

8.2.3   Comparative Analysis of "Discrete Mathematics"

"Discrete Mathematics" (DM) is a theoretical course with 14 students. The course is a second year mandatory course for BSU students in the program of Computer Sciences at the faculty of Physics-Mathematics and Computer Sciences. It was compared with a corresponding course "Algorithm Mathematics" (AM) from Tampere University of Technology (TUT). Outlines of these courses are seen in Table 8.10.

The department of Mathematics is responsible for the course. The staff of this department consist of three full-time professors, six full-time associate professors, four full-time assistant professors and four teachers. The Department of Mathematics conducts the academic process within the frames of the educational programs of the faculty Physics, Mathematics and Computer Sciences as well as other faculties.

The course does not have prerequisite courses. Its size is 5 credits, and it requires on average 125 h of work (25 h for each credit). The credits are divided among different activities as follows: lectures 15 h, tutorials 30 h, homework 80 h. Students should use about 30 h to prepare for their exam.

On the course organizer side, there are 60 contact teaching hours. From one to three teaching assistants work on the course to grade exams and teach tutorials. Unfortunately, there are no computer labs available for use on the course.

There are 14 students on the course. The average number of students that finish the course is 11 (78%). The amount of international students is 0%.

Table 8.10 Outlines of the Discrete Mathematics (BSU) and Algorithm Mathematics (TUT) courses

Course information	BSU	TUT
Bachelor/Master level	Bachelor	Bachelor
Preferred year	1	2
Elective/mandatory	Mandatory	Elective
Number of credits	5	4
Teaching hours	45	49
Preparatory hours	80	65
Teaching assistants	1–3	1–2
Computer labs	No	Yes
Average number of students on the course	14	150
Average pass%	78%	90%
% of international students	No	Less than 5%

The pedagogical comparison of the course shows that the teaching is very much theory based. Professors of the mathematics department are delivering the course, and they do not use modern teaching methods such as blended learning, flipped classroom, project or inquiry based learning etc.

Assessment criteria are as follows: Students are evaluated in a 100-point system, in which 60 points are given on midterm assessments and 40 points on final exams. Midterm assessments include the following components: Work in group (40 points) and two midterm examinations (20 points). Within the frames of 40-point assessment for working in a group, students can be given an abstract/review work for 10 points. Students are obliged to accumulate no less than 11 points in midterm assessments and no less than 21 points at the exams.

Grading follows the same principles as with the course described above.

8.2.3.1 Contents of the Course

The comparison is based on SEFI framework [1]. Prerequisite competencies are presented in Table 8.11. Outcome competencies are given in Table 8.12.

The main differences between Finnish (TUT) and Georgian (BSU) courses are the following: in TUT, teaching is more intense but it covers less topics. The overall hours are somewhat different—TUT has 35 h of lectures and 21 h of tutorials, BSU has 15 lectures and 30 tutorials and TUT does all of this in 7 weeks, whereas BSU uses 15 weeks. TUT also uses plenty of TEL and ICT technologies to support the teaching, BSU does not. Finally, the exams are somewhat different. Finnish students answer in a written exam, BSU students must pass three exams with theoretical questions, but with no proofs.

Table 8.11 Core 0-level prerequisite competencies for DM (BSU) and AM (TUT) courses

Core 0		
Competency	DM (BSU)	AM (TUT)
Arithmetic of real numbers	X	X
Algebraic expressions and formulas	X	X
Linear laws	X	X
Quadratics, cubics, polynomials	X	X
Functions and their inverses	X	X
Sequences, series, binomial expansions	Excl. series	X
Logarithmic and exponential functions	X	X
Proof	X	X
Sets	X	X
Geometry	X	X
Data handling	X	x
Probability	X	

Table 8.12 Core 1-level outcome competencies for DM (BSU) and AM (TUT) courses

Core 1		
Competency	DM (BSU)	AM (TUT)
Sets	X	X
Mathematical logic	X	X
Mathematical induction and recursion	X	X
Graphs	X	
Combinatorics	X	
Simple probability	X	
Probability models	X	

The main drawbacks of the old mathematics syllabi at BSU were that mostly the theoretical mathematical aspects were treated and the corresponding exam lists contained only purely mathematical questions.

Moreover, it should be specially mentioned that the BSU mathematical syllabus "Discrete Mathematics" for Computer Sciences BS programs does not contain the following topics: Binary relation; Boolean algebra; Groups, Rings and Fields; Euclid's division algorithm and Diophantine equations; Coding theory and finite automata; Cryptography. Therefore, modernization of the syllabus "Discrete Mathematics" is very desirable.

BSU will prepare new syllabi for modernized courses that are more in line with European technical university courses. This will be done in order to better prepare the students of BSU for their future careers. However, modernizing courses will not be trivial, since the university has to make up for the different prerequisite skills between European and Georgian enrolling students. The Georgian educational system teaches less mathematics on the high school level, and thus BSU has to design courses that upgrade students' knowledge in these topics as well.

As the overall level of students is sufficiently low, BSU finds that implementing remedial mathematics courses is necessary. This could be done using the Math-Bridge software.

8.3 Analysis of Mathematical Courses in GTU

8.3.1 Georgian Technical University (GTU)

David Natroshvili (✉) and Shota Zazashvili
Georgian Technical University (GTU), Department of Mathematics, Tbilisi, Georgia
e-mail: d.natroshvili@gtu.ge; s.zazashvili@gtu.ge

George Giorgobiani
Georgian Technical University (GTU), Department of Computational Mathematics, Tbilisi, Georgia
e-mail: giorgobiani.g@gtu.ge

Georgian Technical University (GTU) is one of the biggest educational and scientific institutions in Georgia. The overall number of students is approximately 20,000, and there are in total 17 STEM disciplines.

In 1917, the Russian Emperor issued an order to found a polytechnic institute in Tbilisi, the first higher educational institute in the Caucasian region. In 1922, GTU was originally founded as a polytechnic faculty of the Tbilisi State University. Later, in 1928, the departments of the polytechnic faculty merged into an independent institute called Georgian Polytechnic Institute (GPI). GTU continued under this title until 1990, when the institute was granted university status and was renamed Georgian Technical University.

GTU adopted the Bologna process in 2005. Today, the university hosts approximately 20,000 students, 10 faculties and 17 STEM disciplines.

8.3.2   Comparative Analysis of "Mathematics 3"

"Mathematics 3" is a theoretical course with approximately 1500 students. The course is a second year course for engineering students at the university and is a final mathematical mandatory course of engineering programs in GTU. This course was compared with a corresponding course "Engineering Mathematics 4" (EM4) at TUT. The course outlines are seen in Table 8.13.

Department of Mathematics at GTU is responsible for the course. The staff of this department consists of 20 full-time professors, 21 full-time associate professors, 3 full-time assistant professors, 7 teachers, 16 invited professors and 5 technical employees.

Prerequisite courses are Mathematics 1 and 2. The course is a part of the Mathematics course cluster. The course size is 5 credits, and it requires on average 135 h of work (27 h for each credit). The credits are divided among different

Table 8.13 Outlines of Mathematics 3 (GTU) and Engineering Mathematics 4 (TUT) courses

Course information	GTU	TUT
Bachelor/Master level	Bachelor	Bachelor
Preferred year	2	1
Elective/mandatory	Mandatory	Mandatory
Number of credits	5	4
Teaching hours	60	49
Preparatory hours	75	56
Teaching assistants	1–4	1–3
Computer labs	No	Yes
Average number of students on the course	1500	150
Average pass%	75%	85%
% of international students	Less than 1%	Less than 5%

activities as follows: lectures 30 h, tutorials 30 h, homework 75 h, exam 3 h. Students should use about 10 h to prepare for their exam.

On the course organizer side, there are 60 contact teaching hours. From one to four teaching assistants work on the course to grade exams and teach tutorials. Unfortunately, there are no computer labs available for use on the course.

There are approximately 1500 students on the course. The average number of students that finish the course is 1125 (75%). The amount of international students is less than 1%. For this analysis, neither the overall student demographic nor the average rating of the course by the students was available.

The pedagogical comparison of the course shows that the teaching is very much theory based. Professors of the mathematics department are delivering the course, and they do not use modern teaching methods such as blended learning, flipped classroom, project or inquiry based learning etc.

Testing is done on a 100 point scale with 51 points being the passing level. A score between 41 and 50 offers a new attempt at the exam, and every 10 point interval offers a better grade with 91–100 points being the best. Exams are in three different forms: weekly intermediate exams, two midterm exams and a final exam. Testing methods are multiple choice answers or open ended answers done on computers. The final grade is computed by combining the different points in the different tests and normalizing to 100.

"Calculus 2 = Mathematics 3" at GTU does not have any TEL tools available on the course. Therefore, no TEL tools are used to support learning.

8.3.2.1   Contents of the Course

The comparison is based on the SEFI framework [1]. Prerequisite competencies are presented in Tables 8.14 and 8.15. Outcome competencies are given in Tables 8.16, 8.17, and 8.18.

8.3.2.2   Summary of the Results

The main differences between Finnish (TUT) and Georgian (GTU) courses are the following: in TUT, teaching is more intense. The overall hours are quite similar— TUT has 28 h of lectures and 24 h of tutorials, GTU has 30 lectures and 30 tutorials—but TUT does all of this in 7 weeks, whereas GTU uses 15 weeks. TUT also uses plenty of TEL and ICT technologies to support their teaching, GTU does not. Finally, the exams are quite different. Finnish students answer in a pen and paper exam, GTU students must pass three exams (two midterm exams and one final exam) on the computer, with no pure theoretical questions (proofs).

The main drawbacks of the old mathematics syllabi at GTU were that mostly the theoretical mathematical aspects were treated and the corresponding exam lists contained only purely mathematical questions. This means that the application of

Table 8.14 Core 0-level prerequisite competencies of Mathematics 3 (GTU) and EM4 (TUT) courses

Core 0		
Competency	GTU	TUT
Arithmetic of real numbers	X	X
Algebraic expressions and formulas	X	X
Linear laws	X	X
Quadratics, cubics, polynomials	X	X
Functions and their inverses	X	X
Sequences, series, binomial expansions	Excl. series, binomial expansion	X
Logarithmic and exponential functions	X	X
Rates of change and differentiation	X	X
Stationary points, maximum and minimum values	X	X
Indefinite integration	X	X
Proof	X	X
Sets	X	X
Geometry	X	X
Trigonometry	X	X
Co-ordinate geometry	X	X
Trigonometric functions and applications	X	X
Trigonometric identities	X	X

Table 8.15 Core 1-level prerequisite competencies of Mathematics 3 (GTU) and EM4 (TUT) courses

Core 1		
Competency	GTU	TUT
Rational functions	X	X
Complex numbers	X	X
Functions	X	X
Differentiation	X	X
Sequences and series	Excl. series, binomial expansion	X
Vector arithmetic	X	X
Vector algebra and applications	X	X
Matrices and determinants	X	X
Solution of simultaneous linear equations	X	X
Functions of several variables	X	

Table 8.16 Core 0-level outcome competencies of Mathematics 3 (GTU) and EM4 (TUT) courses

Core 0		
Competency	GTU	TUT
Sequences, series, binomial expansions	X	
Indefinite integration	X	X
Definite integration, applications to areas and volumes	X	X

Table 8.17 Core 1-level outcome competencies of Mathematics 3 (GTU) and EM4 (TUT) courses

Core 1		
Competency	GTU	TUT
Sequences and series	X	
Methods of integration	X	X
Applications of integration	X	X

Table 8.18 Core 2-level outcome competencies of Mathematics 3 (GTU) and EM4 (TUT) courses

Core 2		
Competency	GTU	TUT (Tampere)
Ordinary differential equations	X	
First order ordinary differential equations	X	
Second order equations—complementary function and particular integral	X	
Fourier series	X	
Double integrals		X
Further multiple integrals		X
Vector calculus		X
Line and surface integrals, integral theorems		X

taught mathematics had almost no emphasis on the course, which led to lack of motivation in the students.

Moreover, it should be specially mentioned that in the GTU mathematical syllabus "Mathematics 3" (*in Finland* "Calculus 2") for engineering BSc programs, the following topics are not included: Double integrals, Triple integrals, Curvilinear and Surface Integrals, Vector Calculus, Divergence Theorem and Stokes' Theorem. These topics are widely presented in the TUT mathematical curricula.

Due to the above, it seems that an essential modernization of the syllabus Calculus 2 (in GTU Mathematics 3) is very desirable.

At the same time it should be taken into consideration that modernization of a particular syllabus Calculus 2 (in GTU Mathematics 3) will require modification of the syllabuses of prerequisite courses Mathematics 1 and Mathematics 2.

GTU will prepare new syllabi for modernized courses that are more in line with European and American technical university courses. This will be done in order to better prepare the students of GTU for their future careers. However, modernizing courses will not be trivial, since the university has to make up for the different prerequisite skills of European and Georgian enrolling students. The Georgian educational system teaches less mathematics on the high school level, and thus GTU has to design courses that upgrade students' knowledge in these topics as well.

As the overall level of students is sufficiently low, GTU finds that implementing remedial mathematics courses is necessary. This could be done using the Math-Bridge software.

8.3.3 Comparative Analysis of "Probability Theory and Statistics"

"Probability Theory and Statistics" (PTS) is a theoretical course with approximately 200 students. The course is a second year course for engineering students of two departments of two faculties: Faculty of Power Engineering and Telecommunications and Faculty of Informatics and Control Systems. It will be compared with a corresponding course "Probability Calculus" (PC) at Tampere University of Technology (TUT). The outlines of the courses are presented in Table 8.19.

The department responsible for the course is the GTU Department of Mathematics. The staff of this department consists of 20 full-time professors, 21 full-time associate professors, 3 full-time assistant professors, 7 teachers, 16 invited professors and 5 technical employees.

The prerequisite course is Mathematics 2. The course is a part of the Mathematics course cluster. The course size is 5 credits, and it requires on average 135 h of work (27 h for each credit). The credits are divided among different activities as follows: lectures 30 h, tutorials 30 h, homework 75 h, exam 3 h. Students should use about 10 h to prepare for their exam.

On the course organizer side, there are 60 contact teaching hours. From one to two teaching assistants work on the course to grade exams and teach tutorials. Unfortunately, there are no computer labs available for use on the course.

There are approximately 200 students on the course. The average number of students that finish the course is 150 (75%). The amount of international students is less than 1%. For this analysis, neither the overall student demographic nor the average rating of the course by students was available.

The pedagogical comparison of the course shows that the teaching is very much theory based. Professors of mathematics department are delivering the course, and they do not use modern teaching methods such as blended learning, flipped classroom, project or inquiry based learning etc.

Table 8.19 Outlines of probability and statistics courses (PTS) at GTU and (PC) at TUT

Course information	GTU	TUT
Bachelor/Master level	Bachelor	Bachelor
Preferred year	2	2
Elective/mandatory	Mandatory	Elective
Number of credits	5	4
Teaching hours	60	42
Preparatory hours	75	66
Teaching assistants	1–2	1–2
Computer labs	No	Yes
Average number of students on the course	200	200
Average pass%	75%	90%
% of international students	Less than 1%	Less than 5%

Testing is done on a 100 point scale with 51 points being the passing level. A score between 41 and 50 offers a new attempt at the exam, and every 10 point interval offers a better grade with 91–100 being the best. Exams are in three different forms: weekly intermediate exams, two midterm exams and final exam. Testing methods are by multiple choice answers or open ended answers done on computers. The final grade is computed by combining the different points in the different tests and normalizing to 100.

GTU does not have any TEL tools. Therefore, no TEL tools are used to support teaching and learning.

8.3.3.1   Contents of the Course

The comparison is based on the SEFI framework [1]. Prerequisite competencies are presented in Tables 8.20 and 8.21. Outcome competencies are given in Tables 8.22, 8.23, and 8.24.

8.3.3.2   Summary of the Results

The main differences between Finnish (TUT) and Georgian (GTU) courses are the following: in TUT, teaching is more intense. The overall hours also are not similar— TUT has 28 h of lectures and 14 h of tutorials for statistics (7 weeks) and 28 h of

Table 8.20 Core 0-level prerequisite competencies of the probability and statistics courses (PTS) at GTU and (PC) at TUT

Core 0		
Competency	GTU	TUT
Arithmetic of real numbers	X	X
Algebraic expressions and formulas	X	X
Linear laws	X	X
Quadratics, cubics, polynomials	X	X
Functions and their inverses	X	X
Sequences, series, binomial expansions	Excl. series, binomial expansion	X
Logarithmic and exponential functions	X	X
Rates of change and differentiation	X	X
Stationary points, maximum and minimum values	X	X
Indefinite integration	X	X
Proof	X	X
Sets	X	X
Geometry	X	X
Trigonometry	X	X
Co-ordinate geometry	X	X
Trigonometric functions and applications	X	X
Trigonometric identities	X	X

Table 8.21 Core 1-level prerequisite competencies of the probability and statistics courses (PTS) at GTU and (PC) at TUT

Core 1		
Competency	GTU	TUT
Rational functions	X	X
Complex numbers	X	X
Functions	X	X
Differentiation	X	X
Sequences and series	Excl. series	X
Vector arithmetic	X	X
Vector algebra and applications	X	X
Matrices and determinants	X	X
Solution of simultaneous linear equations	X	X
Functions of several variables	X	X

Table 8.22 Core 0-level outcome competencies of the probability and statistics courses (PTS) at GTU and (PC) at TUT

Core 0		
Competency	GTU	TUT
Data handling	X	X
Probability	X	X

Table 8.23 Core 1-level outcome competencies of the probability and statistics courses (PTS) at GTU and (PC) at TUT

Core 1		
Competency	GTU	TUT
Data handling	X	X
Combinatorics	X	X
Simple probability	X	X
Probability models	X	X
Normal distribution	X	X
Sampling	X	
Statistical inference	Excl. issues related to hypothesis testing	

Table 8.24 Core 2-level outcome competencies of the probability and statistics courses (PTS) at GTU and (PC) at TUT

Core 2		
Competency	GTU	TUT
One-dimensional random variables	Excl. Weibull and gamma distributions	X

lectures and 12 h of tutorials for probability (7 weeks), GTU has 30 lectures and 30 tutorials (15 weeks). TUT also uses plenty of TEL and ICT technologies to support their teaching, GTU does not. Finally, the exams are quite different. Finnish students answer in a written exam, GTU students must pass three exams (two midterm exams and one final exam) on the computer, with no pure theoretical questions (proofs).

The main drawbacks of the old mathematics syllabi at GTU were that mostly the theoretical mathematical aspects were treated and the corresponding exam lists contained mostly purely mathematical questions. This means that the application of taught mathematics had almost no emphasis on the course, which led to lack of motivation in the students.

Moreover, it should be specially mentioned that in the GTU syllabus "Probability Theory and Statistics" for engineering BSc programs 12 lectures are devoted to the topics of probability theory, and only the last three lectures to statistics. The following topics are not included: Test of hypothesis; Small sample statistics issues like t-test, F-test, chi-square tests; topics of Analysis of variance; Linear regression etc. Due to the above, it seems that an essential modernization of the syllabus "Probability Theory and Statistics" is very desirable.

8.4   Analysis of Mathematical Courses in UG

Kakhaber Tavzarashvili and Ketevan Kutkhashvilii
University of Georgia (UG), School of IT, Engineering and Mathematics, Tbilisi, Georgia
e-mail: k.tavzarashvili@ug.edu.ge; k.kutkhashvili@ug.edu.ge

The University of Georgia (UG) was founded in 2004. UG is one of the largest private universities in Georgia. Throughout past years, improvement of the quality of education in Georgia has been one of the top priorities and, in this framework, the aim of UG has always been to develop and accomplish high standards as regards academic quality, and as regards student life for the benefit of both Georgian and international students. The University of Georgia is a place that generates and disseminates knowledge. It has created a diverse environment, which forms open-minded and educated persons with human values and the skills necessary to consciously and easily cope with the challenges of the modern world. The academic faculty of the university is represented by a team of creative and enthusiast individuals, willing to educate professionals equipped with the knowledge and skills required in the modern world and ready to make a significant contribution to the welfare of humanity. Today, the University of Georgia is one of the nation's leading universities, with the personal growth and professional development of its students as its main goal. Graduates of the university feel confidence and are ready to enter a competitive job market. Today, the university has the honor to offer its students modern facilities and the learning environment in which they can gain high quality education as well as practical experience. The knowledge and skills acquired at our university are a guarantee of a successful career not only in Georgia, but also in the international labor market. UG is a classical university and it offers a wide range of specializations. From 2014 to 2016 overall numbers of UG students increased up to 50% from 4000 to 6000 students. These students are distributed in six main schools. They are:

- School of Humanities;
- School of Law;
- School of Social Sciences;
- School of IT, Engineering and Mathematics;
- School of Health Sciences and Public Health;
- School of Business, Economics and Management.

The School of IT, Engineering and Mathematics has a STEM profile. There are about 350 students enrolled totally in all academic level. The number of freshmen students for 2016 was 80 students. There are seven different academic programs, four of them on the BSc level, two on the Master level, and the seventh is a PhD level program. For the BSc programs, Informatics has 238 students, Electrical and Computer Engineering has 33, Engineering (for international students) has 10, Mathematics has 7, Computer Science has 181, On the master program, Applied Sciences has 5, and the Exact, Natural and Computer Science program, which is PhD level, has 7 students.

8.4.1   Comparative Analysis of "Precalculus"

Precalculus is an elective course for freshmen students of STEM specializations meant to increase their knowledge and Math skills. This course contains functions, graphs, linear and quadratic functions, inverse, exponential, logarithmic and trigonometric functions; polynomial and rational functions; solving of linear and nonlinear systems of equations and inequalities; sequences and their properties; combinatorial analysis and fundamentals of probability theory, binomial theorem.

Precalculus is a theoretical course but it is based on practical examples. This course was developed for engineering students and all practical examples are connected to real life problems. Earlier it was 26 h in semester and after course modernization practical work hours was added and now it is totally 39 h per semester. There have been 25 students in Precalculus in 2014. Precalculus was compared with "Remedial Instruction" course at Tampere University of Technology (TUT). The course is mandatory for TUT students who do not grasp mathematics well in the beginning of their studies. All the students enrolling at TUT will take a basic skills' test and the weakest 20% of them are directed to "Remedial Instruction". This remedial course is completely done with computers with a tailored introductory course in the Math-Bridge system at TUT. The courses' outlines are seen in Table 8.25.

Table 8.25 Outlines of Precalculus (UG) and Remedial Instruction (TUT) courses

Course information	UG	TUT
Bachelor/Master level	Bachelor	Bachelor
Preferred year	1	1
Elective/mandatory	Elective	Mandatory
Number of credits	6	0
Teaching hours	39	0
Preparatory hours	111	0
Teaching assistants	6	0
Computer labs	No	Yes
Average number of students on the course	200	140
Average pass%	75%	90%
% of international students	Less than 1%	Less than 5%

The Precalculus course was offered for freshmen students from STEM special-izations (Informatics, Electrical and Computer Engineering). Also it was offered to students from Business and Economics specializations.

The Precalculus was prepared by the Department of Mathematics, which is responsible for all mathematics courses in UG. The staff of the Department of Mathematics consists of two full professors, three associated professors, six assistants and invited lecturers (depending of groups). Professors and associated professors are responsible for lectures, while assistants and invited lecturers are responsible for tutorials.

Before course modernization "Precalculus" was mainly a theoretical course. After modernization tutorials should be added and the course should become more applied.

Assessment contains midterm exams (60%) and final examination (40%). Midterm exams contain quizzes (24%) and midexams (36%). Totally, there are eight quizzes (each for 3 points), three midexams (each for 12 points) and final exam (40 points). The minimum number of points required for the final exam is 20. The course is passed when student has more than 50 points and in the final exam more than 20 points.

Each quiz (3 points) contains three tasks from the previous lecture. Each midterm exam (12 points) consists eight tasks including two theoretical questions. The final exam (40 points) consists of 20 tasks including six theoretical questions.

During the lectures professors use various tools for visualization, and presenta-tion tools, in order to present some applications and some dynamic processes, in par-ticular GeoGebra is an illustration tool. There are no mandatory parts of using TEL systems in the exams. We used the Math-Brige system for Pre- and Post-testing.

GeoGebra was used for illustration purposes. It was used to show properties of functions, intersection of functions by axis, finding of intersection points of two functions and so on. The role of TEL systems in the course has been in demonstrating basic mathematics and in visualizing different math topics.

8.4.1.1   Contents of the Course

The comparison is based on the SEFI framework [1]. Prerequisite competencies are presented in Table 8.26. Outcome competencies are given in Table 8.27.

8.4.1.2   Summary of the Results

The University of Georgia has changed the math syllabus in Precalculus. In order to achieve SEFI competencies [1] UG have introduced modern educational technologies in teaching methods. The aim of the modernization was an integration of Math-Bridge and GeoGebra in the study process. UG has done the following modification in curricula and in syllabuses.

Modernization of the syllabus was done since September of 2016 just only in Precalculus. Comparison of syllabuses in other subjects (Precalculus) have shown full compatibility with the subjects of TUT courses. UG have separated Precalculus in two independent courses for STEM and for Business and Economic specializations. In Precalculus for STEM specializations we added some subjects according to the TUT courses content. Also we added 12 h of practical work by using GeoGebra and MATLAB programs.

In 2016 UG has done pre- and post-testing in Precalculus and Calculus 1. Math-Bridge was used for testing. Theoretical and practical examples were prepared in Math-Bridge. Math-Bridge was used as a tool for analyzing the results.

Table 8.26 Core 0-level prerequisite competencies of the Precalculus (UG) and Remedial Instruction (TUT) courses

Core 0		
Competency	UG	TUT
Arithmetic of real numbers	X	X
Algebraic expressions and formulas	X	X
Linear laws	X	X
Quadratics, cubics, polynomials	X	X
Functions and their inverses	X	X

Table 8.27 Core 0-level outcome competencies of the Precalculus (UG) and Remedial Instruction (TUT) courses

Core 0		
Competency	UG	TUT
Graphs	X	
Linear and quadratic functions	X	X
Polynomial and rational functions	X	X
Power functions	X	X
Trigonometric functions	X	X
Sequences, arithmetic and geometric sequences	X	X
Combinatorics and probabilities	X	

8.4.2   Comparative Analysis of "Calculus 1"

Calculus 1 is a mandatory course for students of STEM specializations (Informatics BSc, Electronic and Computer Engineering BSc, Engineering BSc). This course contains the basic properties of inverse, exponential, logarithmic and trigonometric functions; limit, continuity and derivative of a function, evaluating rules of a derivative, function research and curve-sketching techniques, applications of derivative in the optimization problems, L'Hôpital's rule, Newton's method, indefinite and definite integral and their properties, rules of integration, integration of rational functions, evaluating area between curves and surface area using integrals, integrals application in physics, numerical integration.

Calculus 1 is a theoretical course, but it is based on practical examples. This course was developed for engineering students and all practical examples are connected to real life problems. Before the course modernization it was 26 h in a semester and after the course modernization practical work hours were added; now it is totally 39 h per semester. There are 25–50 students on the course each year. It was compared with the corresponding "Engineering Mathematics 1" (EM1) course from Tampere University of Technology (TUT). The course outlines are seen in Table 8.28.

This course was offered to STEM students (Informatics, Electrical and Computer Engineering). Also it was offered to students from Business and Economics specializations with modified content, with more focus on real Business applications.

The "Calculus 1" was prepared by Department of Mathematics, which is responsible for all mathematics courses in UG. The staff of the Department of Mathematics consists of two full professors, three associated professors, six assistants and invited lecturers (depending of groups).

Pedagogy and assessment are done similarly to "Precalculus". During the lectures the professor uses various tools for visualization, presentation tools in order to present some applications and some dynamic processes, in particular GeoGebra

Table 8.28 Outlines of "Calculus 1" (UG) and "Engineering Mathematics 1" (TUT) courses

Course information	UG	TUT
Bachelor/Master level	Bachelor	Bachelor
Preferred year	1	1
Elective/mandatory	Mandatory	Mandatory
Number of credits	6	5
Teaching hours	39	57
Preparatory hours	111	80
Teaching assistants	6	1–3
Computer labs	No	Yes
Average number of students on the course	25	200
Average pass%	80%	90%
% of international students	Less than 1%	Less than 5%

as an illustration tool. There are no mandatory parts of using TEL systems in exams. We have used the Math-Bridge system for Pre- and Post-testing. GeoGebra was used for illustration purposes. It was used to demonstrate the properties of functions, derivatives of a function, applications of derivatives, integrals and applications of integrals.

8.4.2.1   Contents of the Course

The comparison is based on the SEFI framework [1]. Prerequisite competencies are presented in Table 8.29. Outcome competencies are given in Table 8.30.

8.4.2.2   Summary of the Results

UG has changed the Math syllabus in "Calculus 1". In order to achieve SEFI competencies [1] UG has introduced modern educational technologies in teaching methods. The aim of the modernization was an integration of Math-Bridge and GeoGebra in the study process. UG has done the following modifications in curricula and in syllabi.

Modernization of the syllabus was done since September of 2016 just only in "Calculus 1". Comparison of syllabuses in other subjects (Calculus 1 and Calculus 2) have shown full compatibility with the subjects of TUT courses. Tutorials were added with 12 h of practical work by using GeoGebra and MATLAB programs.

Table 8.29 Core 0-level prerequisite competencies of the "Calculus 1" (UG) and "Engineering Mathematics 1" (TUT) courses

Core 0		
Competency	UG	TUT
Arithmetic of real numbers	X	X
Algebraic expressions and formulas	X	X
Linear laws	X	X
Quadratics, cubics, polynomials	X	X
Functions and their properties	X	X

Table 8.30 Core 0-level outcome competencies of the "Calculus 1" (UG) and "Engineering Mathematics 1" (TUT) courses

Core 0		
Competency	UG	TUT
Functions and their basic properties	X	X
Logarithmic and trigonometric functions	X	X
Limits and continuity of a function	X	X
Derivative	X	X
Definite integral and integration methods	X	
Minimum and maxima	X	X
Indefinite integrals	X	

In 2016 UG has done pre- and post-testing in "Precalculus" and "Calculus 1". Math-Bridge was used for testing. Theoretical and practical examples were prepared in Math-Bridge. It was used for analyzing results as well.

Reference

1. SEFI (2013), "A Framework for Mathematics Curricula in Engineering Education" (Eds.) Alpers, B., (Assoc. Eds) Demlova M., Fant C-H., Gustafsson T., Lawson D., Mustoe L., Olsson-Lehtonen B., Robinson C., Velichova D. (http://www.sefi.be).

Chapter 9
Case Studies of Math Education for STEM in Armenia

9.1 Analysis of Mathematical Courses in ASPU

Lusine Ghulghazaryan and Gurgen Asatryan
Faculty of Mathematics, Physics and Informatics, Armenian State Pedagogical University (ASPU), Yerevan, Armenia
e-mail: lusina@mail.ru

9.1.1 Armenian State Pedagogical University (ASPU)

Armenian State Pedagogical University (ASPU) was established on November 7, 1922 and in 1948 it was named after the great Armenian Enlightener Educator Khachatur Abovian. ASPU implements a three-level education system (Bachelor, Master and Doctorate studies). It has ten faculties.

The Faculty of Mathematics, Physics and Informatics was formed in 2012 by merging the two Faculties of "Mathematics and Informatics" and "Physics and Technology". The faculty operates under a two-level education system. It offers Bachelor's degree courses (4 years) and Master's degree courses (2 years). The faculty prepares teachers in the specializations of Mathematics, Physics, Natural Sciences, Informatics and Technology.

The faculty has four Departments: the Department of Mathematics and its Teaching Methodology, the Department of Physics and its Teaching Methodology, the Department of Technological Education, and the Department of Informatics and its Teaching Methodology.

© The Author(s) 2018   169
S. Pohjolainen et al. (eds.), *Modern Mathematics Education for Engineering Curricula in Europe*, https://doi.org/10.1007/978-3-319-71416-5_9

The Department of Mathematics and its Teaching Methodology was formed in 2016 by merging three Departments of Higher Algebra and Geometry, Mathematical Analysis and Theory of Functions, Teaching Methodology of Mathematics.

9.1.2 Comparative Analysis of "Linear Algebra and Analytic Geometry"

The course is given for the first year Bachelor students of specializations Informatics and Physics. For the students specialized in Mathematics there are two separate courses—"Linear Algebra" and "Analytic Geometry" (LA&AG). The average number of students is 35 in each specialization. The course "Linear Algebra and Analytic Geometry" was compared with "Engineering Mathematics 1" (EM2) course at Tampere University of Technology; see Table 9.1 for course outlines.

The Department of Mathematics and its Teaching Methodology is responsible for the course. The course is taught 4 h per week—2 h for lectures and 2 h for practice. The average number of students is 35 in each specialization (Mathematics, Physics, Informatics); most of them are female. There are no international students in the Faculty. Students use the TEL-systems in the computer laboratory created by Tempus MathGeAr project. In particular they use the Math-Bridge system during their study.

9.1.2.1 Contents of the Course

The course is dedicated to working with matrices, systems of linear equations, vector spaces and subspaces, linear mapping, coordinate method, equations of lines and planes, second order curves and surfaces.

Table 9.1 Outlines of LA&AG (ASPU) and EM2 (TUT) courses

Course information	ASPU	TUT
Bachelor/Master level	Bachelor	Bachelor
Preferred year	1	1
Selective/mandatory	Mandatory	Mandatory
Number of credits	5	5
Teaching hours	64	57
Preparatory hours	86	76
Teaching assistants	1	1–3
Computer labs	Available	Available
Average number of students on the course	35	200
Average pass%	70%	90%
% of international students	0%	Less than 5%

The list of contents is:

1. Vector spaces, subspaces, linear mappings.
2. Complex numbers, the module and the argument of a complex number.
3. Matrix Algebra, determinant, rank, inverse of the matrix.
4. Systems of linear equations, Gauss method, Cramer's rule, Kronecker–Capelli's theorem, systems of linear homogeneous equations.
5. Linear mappings, the rank and defect of a linear mapping, the kernel of a linear mapping
6. Polynomials, the roots of a polynomial, Bézout's theorem, the Horner scheme.
7. Coordinate method, distance of two points, equations of lines and planes, distance of a point and a line or plane.
8. Second order curves and surfaces.

Prerequisites for the course is knowledge in elementary mathematics. Outcome courses are General Algebra, Theory of Topology and Differential Geometry.

The objectives for the course are: To provide students with a good understanding of the concepts and methods of linear algebra; to help the students develop the ability to solve problems using linear algebra; to connect linear algebra to other fields of mathematics; to develop abstract and critical reasoning by studying logical proofs and the axiomatic method.

The assessment is based on four components, two midterm examinations, the final examination and the attendance of the student during the semester. The points of the students are calculated by $m = (1/4)a + (1/4)b + (2/5)c + (1/10)d$, where a, b, c, d are the points of intermediate examinations, final examination and the point of attendance, respectively (maximal point of each of a, b, c, d is 100). The satisfactory point starts from 60.

9.1.2.2   Course Comparison Within SEFI Framework

The comparison is based on the SEFI framework [1]. Prerequisite competencies are presented in Table 9.2. Outcome competencies are given in Tables 9.3 and 9.4.

Table 9.2 Core 0-level prerequisite competencies of LA&AG (ASPU) and EM2 (TUT) courses

Core 0		
Competency	ASPU	TUT
Arithmetic of real numbers	X	X
Algebraic expressions and formulas	X	X
Linear laws	Excl.: interpret simultaneous linear inequalities in terms of regions in the plane	X
Quadratics, cubics, polynomials	X	X
Geometry	X	X

Table 9.3 Core 0-level
outcome competencies of
LA&AG (ASPU) and EM2
(TUT) courses

Core 0		
Competency	ASPU	TUT
Linear laws	X	X
Coordinate geometry	X	X

Table 9.4 Core 1-level
outcome competencies of
LA&AG (ASPU) and EM2
(TUT) courses

Core 1		
Competency	ASPU	TUT
Vector arithmetic	X	X
Vector algebra and applications	X	X
Matrices and determinants	X	X
Solution of simultaneous linear equations	X	X
Linear spaces and transformations	X	
Conic sections	X	

9.1.2.3   Summary of the Results

The course "Linear Algebra and Analytic Geometry" in ASPU has been compared
with the corresponding course "Engineering Mathematics 2" at Tampere University
of Technology (TUT).

The teaching procedure in ASPU is quite theoretical and the assessment is
based mainly on the ability of students of proving the fundamental theorems. This
theorem-to-proof method of teaching is quite theoretical, while the corresponding
courses in TUT are quite practical and applicable. In this regard, the assessment in
TUT is based on the emphasis on the ability of applying the fundamental theorems
in solving problems.

According to the analysis of the teaching methodology, the course "Linear
Algebra and Analytic Geometry" in ASPU should be reorganized so that the modern
aspects of the case can be presented more visually. In particular during the teaching
process some applications of the case should be provided. Some of the main
theorems and algorithms should be accompanied by programming in MATLAB.

The main steps of modernization to be taken are:

- Include more practical assignments.
- Demonstrate applications of algebra and geometry.
- Construct bridge between the problems of linear algebra and programming (this
  would be quite important especially for students of Informatics).
- Use ICT tools for complex calculations.
- Implement algorithms for the basic problems of linear algebra. (Gauss method,
  Cramer's rule, matrix inversion...)
- Use Math-Bridge for more practice and for the theoretical background.
- Students should do experiments for geometric objects using ICT tools.
- Before the final examination students should prepare a paperwork with the
  solution of problems assigned by the teacher and the results of their experiments.

9.1.3 Comparative Analysis of "Calculus 1"

The course is given for one and a half year Bachelor students of specializations Informatics and Physics. For the students in the specialization Mathematics is given 2 years. Average number of students is 35 (in each specialization). The course was compared with the corresponding course "Engineering Mathematics 1" (EM1) at Tampere University of Technology (TUT). The course outlines are presented below; see Table 9.5.

The Department of Mathematics and its Teaching Methodology is responsible for the course. The course is taught 4 h per week—2 h for lectures and 2 h for practice. The average number of students is 35 in each specialization (Mathematics, Physics, Informatics), most of them are female. There are no international students in the Faculty. Students use the TEL-systems in a computer laboratory created by Tempus MathGeAr project. In particular they use the Math-Bridge system during their study.

9.1.3.1 Contents of the Course

The course is dedicated to working with the rational and real numbers, limits of numerical sequences, limit of functions, continuity of function, monotonicity of function, derivative, differential of function, convexity and concavity of the graph of the function, investigation of function and plotting of graphs.

The list of contents is:

1. The infinite decimal fractions and set of real number.
2. Convergence of the numerical sequences.
3. Limit of a function and continuity of function.
4. Derivative and differential of function.
5. Derivatives and differentials of higher orders, Taylor's formula.

Table 9.5 Outlines of Calculus 1 (ASPU) and EM1 (TUT) courses

Course information	ASPU	TUT
Bachelor/Master level	Bachelor	Bachelor
Preferred year	1	1
Selective/mandatory	Mandatory	Mandatory
Number of credits	5	5
Teaching hours	64	57
Preparatory hours	86	76
Teaching assistants	1	1–3
Computer labs	Available	Available
Average number of students on the course	35	200
Average pass%	70%	90%
% of international students	0%	Less than 5%

6. Monotonicity of a function.
7. Extremes of a function.
8. Convexity and concavity of the graph of the function.
9. Investigation of a function and plotting of graphs.

A prerequisite for the course is knowledge of elementary mathematics. Outcome courses are Functional analysis, Differential equations and Math-Phys. equations. Objectives of the course for the students are:

- To provide students with a good understanding of the fundamental concepts and methods of Mathematical Analysis.
- To develop logical reasoning, provide direct proofs, proofs by contradiction and proofs by induction.
- To teach students to use basic set theory to present formal proofs of mathematical statements.
- To develop the ability of identifying the properties of functions and presenting formal arguments to justify their claims.

The assessment is based on three components, two midterm examinations and the attendance of the student during the semester. The points of the students are calculated by $m = (2/5)a + (2/5)b + (1/5)c$, where a, b, c are the points of midterm examinations and the points of attendance, respectively (maximal number of points of each of a, b, c is 100). The satisfactory number of points starts from 60.

9.1.3.2 Course Comparison Within SEFI Framework

The comparison is based on the SEFI framework [1]. Prerequisite competencies are presented in Table 9.6. Outcome competencies are given in Tables 9.7 and 9.8.

Table 9.6 Core 0-level prerequisite competencies of "Calculus 1" (ASPU) and EM1 (TUT) courses

Core 0		
Competency	ASPU	TUT
Functions and their inverses	X	X
Sequences, series, binomial expansions	Excl.[a]	X
Logarithmic and exponential functions	X	X
Rates of change and differentiation	X	X
Stationary points, maximum and minimum values	Excl.[b]	X

[a]Obtain the binomial expansions of $(a + b)^2$ for a rational number; use the binomial expansion to obtain approximations to simple rational functions
[b]Obtain the second derived function of simple functions; classify stationary points using second derivative

Table 9.7 Core 1-level outcome competencies of "Calculus 1" (ASPU) and EM1 (TUT) courses

Core 0		
Competency	ASPU	TUT
Sequences, series, binomial expansions	X	
Stationary points, maximum and minimum values	X	X
Proof	X	X

Table 9.8 Core 1-level outcome competencies of "Calculus 1" (ASPU) and EM1 (TUT) courses

Core 1		
Competency	ASPU	TUT
Hyperbolic functions	X	X
Rational functions	X	X
Functions	X	X
Differentiation	X	X

9.1.3.3 Summary of the Results

The course "Calculus 1" in ASPU has been compared with the corresponding course "Engineering Mathematics 1" at Tampere University of Technology (TUT).

The teaching procedure in ASPU is quite theoretical and the assessment is based mainly on the ability of students of proving the fundamental theorems. This theorem-to-proof method of teaching is quite theoretical, while the corresponding courses in TUT are quite practical and applicable. In this, the assessment in TUT is based on the emphasis on the ability of applying the fundamental theorems in solving problems.

According to the analysis of the teaching methodology, the course "Calculus 1" in ASPU should be reorganized so that the modern aspects of the case be presented more visually. In particular during the teaching process some applications of the case should be provided. Some of the main concepts and properties should be accompanied by programming in Wolfram Mathematica.

The main steps of modernization are:

- Include more practical assignments.
- Demonstrate applications of calculus.
- Construct a bridge between the problems of calculus and physical phenomena (for students of physics) and programming (for students of Informatics).
- Use ICT tools for complex calculations.
- Use Math-Bridge for more practice and for the theoretical background.
- Students should do experiments for graph of function using ICT tools.
- Before the final examination students should prepare a paperwork with the solution of problems assigned by the teacher and the results of their experiments.

9.2   Analysis of Mathematical Courses in NPUA

Ishkhan Hovhannisyan (✉) and Armenak Babayan
Faculty of Applied Mathematics and Physics, National Polytechnic University of
Armenia (NPUA), Yerevan, Armenia

9.2.1   National Polytechnic University of Armenia (NPUA)

National Polytechnic University of Armenia (NPUA) is the legal successor of Yerevan Polytechnic Institute, which was founded in 1933, having only 2 departments and 107 students. The institute grew along with the Republic's industrialization and in 1980–1985 reached its peak with about 25,000 students and more than 66 majors, becoming the largest higher education institution in Armenia and one of the most advanced engineering schools in the USSR. On November 29, 1991, the Yerevan Polytechnic Institute was reorganized and renamed State Engineering University of Armenia (SEUA). In 2014, by the Resolution of the Government of the Republic of Armenia (RA) the traditional name "Polytechnic" was returned to the University and SEUA has been reorganized and renamed to National Polytechnic University of Armenia.

During 83 years of its existence, the University has produced nearly 120,000 graduates, who have contributed greatly to the development of industry, forming a powerful engineering manpower and technology base for Armenia. At present NPUA has about 9000 students. The great majority of them are STEM students. The number of the regular academic staff of the University exceeds 800, most of them with Degrees of Candidate or Doctor of Sciences. With its developed research system and infrastructure the University is nationally recognized as the leading center in technical sciences.

Today, at its central campus located in Yerevan and the Branch Campuses—in Gyumri, Vanadzor and Kapan, the University accomplishes four study programs of vocational, higher and post-graduate professional education, conferring the qualification degrees of junior specialist, Bachelor, Master and researcher. Besides, the degree programs, the University also offers extended educational courses by means of its faculties and a network of continuing education structures. The scope of specialization of the University includes all main areas of engineering and technologies represented by 43 Bachelor's and 26 Master's specializations in Engineering, Industrial Economics, Engineering Management, Applied Mathematics, Sociology and others, offered by 12 faculties. Totally there are more than 40 STEM disciplines in NPUA.

Apart from the faculties, NPUA has a Foreign Students Division which organizes the education of international students from across the Middle East, Asia and Eastern Europe. Their overall number today is almost 200. The languages of instruction are Armenian and English.

The University has a leading role in reforming the higher education system in Armenia. NPUA was the first higher education institute (HEI) in RA that introduced two- and three-level higher education systems, and it implemented the European Credit Transfer System (ECTS) in accordance with the developments of the Bologna Process.

During the last decade, the University has also developed an extended network of international cooperation including many leading Universities and research centers of the world. The University is a member of European University Association (EUA), Mediterranean Universities Network, and Black Sea Universities Network. It is also involved in many European and other international academic and research programs. The University aspires to become an institution, where the education and educational resources are accessible to diverse social and age groups of learners, to both local and international students, as well as to become an institution which is guided by global perspective and moves toward internationalization and European integration of its educational and research systems.

9.2.2   Comparative Analysis of "Mathematical Analysis-1"

Mathematical Analysis-1 (MA-1) is a fundamental mathematical discipline for major profile Informatics and Applied Mathematics (IAM). It is mainly a theoretical course, but it also contains certain engineering applications, such as derivatives arising from engineering and physics problems. There are about 50 first year students (four of which are international) at IAM and they all study this course. "Mathematical Analysis-1" was compared with a similar course "Engineering Mathematics 1" from Tampere University of Technology (TUT). The course outlines are seen in Table 9.9.

Table 9.9 Outlines of MA-1 (NPUA) and EM1 (TUT) courses

Course information	NPUA	TUT
Bachelor/Master level	Bachelor	Bachelor
Preferred year	1	1
Selective/mandatory	Mandatory	Mandatory
Number of credits	6	5
Teaching hours	80	57
Preparatory hours	80	76
Teaching assistants	1–2	1–3
Computer labs	Yes	Yes
Average number of students on the course	50	200
Average pass%	75%	90%
% of international students	Less than 8%	Less than 5%
Description of groups	50 students in two groups, 25 in each. Avg. age is 17 years. Male students twice more than females	

The prerequisite for Mathematical Analysis-1 is high school mathematics. Mathematical Analysis-1 is fundamental for all Mathematical disciplines. The course of Mathematical Analysis 1 together with Mathematical Analysis-2,-3, Linear Algebra and Analytical Geometry, and others is included in the group of mandatory mathematical courses. This group is a requirement for all Bachelor level students of IAM during first year of study.

The chair of General Mathematical Education is responsible for this course for IAM-profile. There are 2 full professors and 25 associate professors working at the chair. The total number of credits is 6. It is an 80-h course, including 32 h of lectures and 48 h of tutorials.

9.2.2.1   Teaching Aspects

The course of Mathematical Analysis-1 is established for the first year students and is quite theoretical. So the pedagogy is traditional: students listen to lectures, accomplish some tasks during tutorials and do their homework. Project-based learning is used in this course too, which makes learning process more interesting, sometimes funny and even competitive. Sometimes the group of students is divided into several subgroups and every subgroup fulfills some task. This kind of work in subgroups is very competitive and students like it. Some teachers use Moodle for distance learning.

NPUA uses the following rating system. The maximum grade is 100 points; one can get 50 points during the semester and another 50 points (as a maximum) is left for the final exam. During the semester students get their 10 points for work in the class and 20 points for each of two midterm tests. These tests allow the teacher to assess the students' work during the semester. Exams are either in oral or written form and include theoretical questions (e.g. a theorem with a proof) and computational tasks. The final grade is the sum of the semester and exam grades. A final grade of at least 81 corresponds to "excellent" (ECTS grade A); a grade from 61 to 80 corresponds to "good" (ECTS grade B); if the sum is between 40 and 60, the student's grade is "satisfactory" (ECTS grade C). Finally, students fail (grade "non-satisfactory", equivalent to ECTS grade F), if their final grade is less than 40.

There is a 2-h lecture on Mathematical Analysis-1 and a 3-h tutorial every week. During tutorials students solve problems (complete computational tasks) under teacher's direction. Students may be given home tasks, which must be done during preparatory hours. Computer labs are not used for every tutorial, but the computers are used to control and grade programming homework.

9.2.2.2   Use of Technology

Some programming languages (C++ or Pascal) are used for homework, writing of programs (topics are synchronized with the course content) is a mandatory part of the midterm tests. E-mail and social networks are sometimes used to have closer connection with students, give assignments etc. After participation in the MathGeAr project we use Math-Bridge and Moodle for teaching Mathematical Analysis-1.

There are about 50 IAM students attending the course; for these profiles lectures and tutorials are set separately. Currently it is still too early to give details of the course outcome; we will have the results after the second middle test and final examination. But student's unofficial feedback is very positive.

Finally, we would like to mention that quite recently we have got four foreign students in this course, and so far, they appreciate the course Mathematical Analysis-1.

9.2.2.3 Course Comparison Within SEFI Framework

The comparison is based on the SEFI framework [1]. Prerequisite competencies are presented in Table 9.10. Outcome competencies are given in Tables 9.11 and 9.12.

Table 9.10 Core 0-level prerequisite competencies for MA-1 (NPUA) and EM1 (TUT) courses

Core 0		
Competency	NPUA	TUT
Arithmetic of real numbers	X	X
Algebraic expressions and formulas	X	X
Linear laws	X	X
Quadratics, cubics, polynomials	Exl. derivative	X
Functions and their inverses	Exl. the limit of a function	X
Sequences, series, binomial expansions	Exl. binomial expansions	X
Logarithmic and exponential functions	X	X
Proof	X	
Geometry	X	X
Trigonometry	X	X
Coordinate geometry	X	X
Trigonometric functions and applications	X	X
Trigonometric identities	X	X

Table 9.11 Core 0-level outcome competencies for MA-1 (NPUA) and EM1 (TUT) courses

Core 0		
Competency	NPUA	TUT
Rates of change and differentiation	X	X
Stationary points, maximum and minimum values	X	X
Functions of one variable	X	X

Table 9.12 Core 1-level outcome competencies for MA-1 (NPUA) and EM1 (TUT) courses

Core 1		
Competency	NPUA	TUT
Functions	Exl. partial derivatives	Exl. partial derivatives
Differentiation	X	X
Sequences and series	Exl. series	
Mathematical induction and recursion	X	X

9.2.2.4   Summary of the Results

One of the purposes of TEMPUS MathGeAr-project was to modernize selected national math courses meeting SEFI criteria after comparing these with the corresponding EU-courses. The SEFI framework for math curricula in STEM education [1] provides the following list of competencies:

- thinking mathematically,
- reasoning mathematically,
- posing and solving mathematical problems,
- modeling mathematically,
- representing mathematical entities,
- handling mathematical symbols and formalism,
- communicating in with and about mathematics,
- making use of aids and tools.

After studying SEFI framework and comparing national math curricula with those of EU considerable commonality around the aims and objectives, curriculum content and progression, and aspirations for problem-solving are revealed. The mathematics expectations at NPUA selected course are comparable to those at TUT. But there are some remarkable distinctive features, discussed in comparative analysis. The general feeling of EU experts at TUT (Tampere) and UCBL (Lyon) is that the course in the partner universities could have a more applied nature and in the course learning technology could be better used. In order to make the NPUA math curricula converge to the European standards, thus ensuring transferability of learning results and introducing best European educational technologies for mathematics, following recommendations of EU experts, the NPUA implemented the following to the curriculum:

- changed syllabus (contents and the way of presentation, "theorem-to-proof" style was modified by putting more emphasis on applications);
- added more topics, applications and examples related to the engineering disciplines;
- started using mathematical tool programs (MATLAB, Scilab, R, etc.);
- started using Math-Bridge for the Mathematical Analysis-1 course;
- added minor student project tasks to the course, including using web resources.

9.2.3   Comparative Analysis of "Probability and Mathematical Statistics"

This course is one of the courses in the program "Informatics and Applied Mathematics". This program is applied because it prepares specialists in IT technologies, mainly programmers, specialists in computer sciences, financial markets experts and so on. But for this kind of work solid mathematical knowledge is necessary, so rigorous theoretical facts are an essential part of the course. The number of students of our faculty is approximately 300, and every year approximately 50 students enroll the course "Probability and Mathematical Statistics". Mathematics is an essential part in the study program, almost all courses of the program are connected with mathematics, or need solid mathematical background, because the ability of thinking mathematically and the ability to create and use mathematical models are the most important acquirements that our graduates must have. The course "Probability and Mathematical Statistics" was compared with a similar course on "Probability Calculus" at Tampere University of Technology (TUT). The course outlines are seen in Table 9.13.

This course is one of the four most important courses of the program, because probabilistic thinking is one of the most important abilities for a modern specialist. It starts in the fourth semester (last semester of the second year) and the duration of this course is 1 year. Prerequisite courses are the general courses of Mathematical Analysis, Linear Algebra and Analytic Geometry. This course was selected from the cluster of mathematical courses mandatory for the students of the "Informatics" specialization. Topics of the course are used in following courses: "Numerical Methods" "Mathematical Physics Equations" and this course is the base for the Master's program courses "Mathematical Statistics", "Stochastic Processes", "Information Theory".

The Chair of Specialized Mathematical Education is responsible for the course. This chair is responsible for the mathematical courses of the Master programs of NPUA and all courses (in Bachelor and Master programs) for "Informatics and

Table 9.13 Outlines of the "Probability and Mathematical Statistics" (NPUA) and "Probability Calculus" (TUT) courses

Course information	NPUA	TUT
Bachelor/Master level	Bachelor	Bachelor
Preferred year	2	2
Selective/mandatory	Mandatory	Selective
Number of credits	6	4
Teaching hours	52	42
Preparatory hours	52	66
Teaching assistants	–	1–2
Computer labs	2	Available
Average number of students on the course	50	200
Average pass%	80%	90%
% of international students	None	Less than 5%

Applied Mathematics" speciality. There are five full professors and eight associate professors working at the Chair.

The number of the credits for the course "Probability and Mathematical Statistics" is six. Two of them are for the lectures, and four for the practical work.

Teaching hours are three hours per week. One hour for lecture and 2 h for practical work. No preparatory hours are planned, because it is supposed that students have enough mathematical knowledge (general courses of mathematical analysis, linear algebra and analytic geometry). Two computer laboratories help us organize teaching process effectively.

The average number of students in the course is 50. Approximately 80% successfully finish it. This course is delivered in Armenian, so foreign students may appear in the group only occasionally. But the same course may be delivered in English for foreign students.

9.2.3.1   Course Comparison Within SEFI Framework

The comparison is based on the SEFI framework [1]. Prerequisite competencies are presented in Tables 9.14 and 9.15. Outcome competencies are given in Tables 9.16 and 9.17.

Table 9.14 Core 0-level prerequisite competencies of the "Probability and Mathematical Statistics" (NPUA) and "Probability Calculus" (TUT) courses

Core 0		
Competency	NPUA	TUT
Arithmetic of real numbers	X	X
Algebraic expressions and formulas	X	X
Linear laws	X	X
Quadratics, cubics, polynomials	X	X
Functions and their inverses	X	X
Analysis and calculus	X	X
Sets	X	X
Geometry and trigonometry	X	X

Table 9.15 Core 1-level prerequisite competencies of the "Probability and Mathematical Statistics" (NPUA) and "Probability Calculus" (TUT) courses

Core 1		
Competency	NPUA	TUT
Hyperbolic and rational functions	X	X
Functions	X	X
Differentiation	X	X
Methods of integration	X	X
Sets	X	X
Mathematical induction and recursion	X	X
Matrices and determinants	X	X
Least squares curve fitting	X	X
Linear spaces and transformations	X	X

Table 9.16 Core 0-level outcome competencies of the "Probability and Mathematical Statistics" (NPUA) and "Probability Calculus" (TUT) courses

Core 0		
Competency	NPUA	TUT
Data handling	X	X
Probability	X	X

Table 9.17 Core 1-level outcome competencies of the "Probability and Mathematical Statistics" (NPUA) and "Probability Calculus" (TUT) courses

Core 1		
Competency	NPUA	TUT (Tampere)
Data handling	X	X
Combinatorics	X	X
Simple probability	X	X
Probability models	X	X
Normal distribution	X	X
Statistical inference	Exc. advanced part of hypothesis testing	X

9.2.3.2   Summary of the Results

The general feeling of EU experts at TUT and Lyon is that the course could have a more applied nature and the learning technology could be better used. In order to make the NPUA math curricula converge to the European standards, thus ensuring transferability of learning results and introducing the best European educational technologies for mathematics, following recommendations of EU experts, the NPUA implemented the following to its curriculum: one

- changed syllabus (contents and the way of presentation, "theorem-to-proof" style was modified by putting more emphasis on applications);
- added more topics, applications and examples related to the engineering disciplines;
- started using mathematical tool programs (MATLAB, Scilab, R, etc.);
- started using Math-Bridge for the course;
- added minor student project tasks to the course, including using web resources.

Reference

1. SEFI (2013), "A Framework for Mathematics Curricula in Engineering Education" (Eds.) Alpers, B., (Assoc. Eds) Demlova M., Fant C-H., Gustafsson T., Lawson D., Mustoe L., Olsson-Lehtonen B., Robinson C., Velichova D. (http://www.sefi.be).

Chapter 10
Overview of the Results and Recommendations

Sergey Sosnovsky, Christian Mercat, and Seppo Pohjolainen

10.1 Introduction

The two EU Tempus-IV projects MetaMath (www.metamath.eu) and MathGeAr (www.mathgear.eu) have brought together mathematics educators, TEL specialists and experts in education quality assurance from 21 organizations across six countries. A comprehensive comparative analysis of the entire spectrum of math courses in the EU, Russia, Georgia and Armenia has been conducted. Its results allowed the consortium to pinpoint and introduce several curricular modifications while preserving the overall strong state of the university math education in these countries. The methodology, the procedure and the results of this analysis are presented here.

During the first project year 2014 three international workshops were organized in Tampere (TUT), Saarbrucken (DFKI & USAAR) and Lyon (UCBL), respectively. In addition, national workshops were organized in Russia, Georgia and Armenia. The purpose of the workshops was to get acquainted with engineering mathematics curricula in the EU and partner countries, with teaching and learning methods used in engineering mathematics, as well as the use of technology in instruction of mathematics. Finally, an evaluation methodology was set up for degree and course comparison and development.

S. Sosnovsky
Utrecht University, Utrecht, the Netherlands
e-mail: s.a.sosnovsky@uu.nl

C. Mercat
IREM Lyon, Université Claude Bernard Lyon 1 (UCBL), Villeurbanne, France
e-mail: christian.mercat@math.univ-lyon1.fr

S. Pohjolainen (✉)
Tampere University of Technology (TUT), Laboratory of Mathematics, Tampere, Finland
e-mail: seppo.pohjolainen@tut.fi

© The Author(s) 2018

S. Pohjolainen et al. (eds.), *Modern Mathematics Education for Engineering Curricula in Europe*, https://doi.org/10.1007/978-3-319-71416-5_10

185

10.2 Curricula and Course Comparison

To accomplish curricula comparison and to set up guidelines for further development SEFI framework [1] was used from the beginning of the project. The main message of SEFI is that in evaluating educational processes we should shift from contents to competences. Roughly said, competences are the knowledge and skills students have when they have passed the courses and this reflects not only on the course contents but especially on the teaching and learning processes, use of technology and assessment of the learning results. SEFI presents recommendations on the topics that BSc engineering mathematics curricula in different phases or levels of studies should contain. The levels are Core 0—prerequisite mathematics, Core 1—common contents for most engineering curricula, Core 2—elective courses to complete mathematics education on chosen engineering area and finally Core 3 for advanced mathematical courses. SEFI makes recommendations on assessing students knowledge and on the use of technology. Here it is also emphasized that learning should be meaningful from the students' perspective. Students' perceptions on mathematics in each of the universities were investigated and results were presented in Chap. 1, Sect. 1.3.

The SEFI framework and related EU-pedagogy was described in Chap. 1, Sect. 1.1. To make a comparison of curricula and courses, a methodology was created. This methodology was described in detail in Chap. 2. Comprehensive information on partner countries' curricula, courses and instruction was collected and organized in the form of a database. For comparison similar data was collected from participating EU-universities, Tampere University of Technology (TUT), Finland and university of Lyon (UBCL), France. This data includes:

- **University:** university type, number of students, percentage engineering students, number of engineering disciplines, degree in credits, percentage of math in degree.
- **Teaching:** Teacher qualifications, delivery method, pedagogy, assessment, SEFI depth aim, modern lecture technology, assignment types, use of third party material, supportive teaching.
- **Selected course details:** BSc or MSc level, preferred year, selective/mandatory, prerequisite courses, outcome courses, department responsible, teacher position, content, learning outcomes, SEFI level, credits, duration, student hours (and their division), average number of students.
- **Use of ICT/TEL:** Tools used, mandatory/extra credit, optional, e-learning/blended/traditional, Math-Bridge, calculators, mobile technology.
- **Resources:** Teaching hours, assistants, computer labs, average amount of students in lectures/tutorials, use of math software, amount of tutorial groups, access to online material.

10.3 Comparison of Engineering Curricula

The overall evaluation of partner universities curricula shows that they aim to cover the SEFI core content areas, especially Cores 0 and 1 of BSc-level engineering mathematics.

Considering only coverage may, however, lead to erroneous conclusions, if the amount of mandatory ECTS in mathematics is not considered simultaneously. For example, one university may have one 5 ECTS mathematics course during a semester, while the second has two 5 ECTS courses, which cover the same topics in the same time. In this case much more time is allocated in the second university to study the same topics. This means that the contents will be studied more thoroughly, students can use more time for their studies, and the learning results will better.

The ECTS itself are comparable, except in Russia, where 1 Russian credit unit corresponds to 36 student hours compared with 25–30 h per ECTS in other universities. The amount of mandatory ECTS and contact hours used in teaching depends on the policy of the university and it varies more as is seen in Tables 10.1, 10.2, and 10.3 below.

To compare curricula, the following information was collected from EU and partner universities. Table 10.1 shows the amount mandatory mathematics in Engineering BSc programs in Finland (TUT), France (UCBL) and Russia (OMSU); the corresponding information for Armenia (ASPU), (NPUA), and Georgia (ATSU), (BSU), (GTU), (UG), is given in Tables 10.2 and 10.3. The first column presents the country and partner university. The second column shows one ECTS as the hours a student should work for it. It covers both contact hours (lectures, exercise classes, etc.) and independent work (homework, project work, preparation for exams etc.). The third column shows one ECTS as contact hours like lectures, exercise classes etc., where the teacher is present. The fourth column shows the mandatory amount of mathematics in BSc programs, and the fifth column shows all (planned) hours a student should use to study mathematics. It has been calculated as the product of mandatory ECTS and hours/ECTS for each university/BSc program.

In the first table, figures from Finland (TUT), France (UCBL), and Russia (OMSU) are given. As the educational policy in Russia is determined on the national level, the numbers from other Russian partner universities are very much alike. That is why we have only the Ogarev Mordovia State University (OMSU) representing the Russian universities for comparison.

Table 10.1 Mandatory mathematics in EU- and Russian engineering BSc programs

Country	One ECTS as student hours	One ECTS as contact hours	BSc program	Mathematics in ECTS	Mathematics in student hours	Notes
Finland (TUT)	26.67	11–13	All (except natural sciences)	27	720	Additional elective courses can be chosen
Finland (TUT)	26.67	11–13	Natural sciences	60	1600	Mathematics major
France (UCBL)	25–30	10	Generic	48	1200–1440	First 2 years of study
France (GPCE)			Generic		864	Higher School Preparatory Classes
Russia (OMSU)	1 CU = 36h ≈ 1.33 ECTS	1CU = 18 contact hours	16 technical and engineering BSc programs	7–75 CU (9–100 ECTS)	252–2700	Russian CU = 36 h. Half of the programs have more than 20 CU (27 ECTS), half less than 20 CU (27 ECTS). Exploitation of Transport and Technological Machines and Complexes BSc has only 7 CU, Fundamental Informatics and Information Technologies has 75 CU
Russia (OMSU)	1 CU = 36h	1CU = 18 contact hours	Informatics and Computer Science	35 CU (47 ECTS)	1260	Russian CU = 36 h. An example of a BSc program in OMSU

Table 10.2 Mandatory mathematics in Armenian engineering BSc programs

Country	One ECTS as student hours	One ECTS as contact hours	BSc program	Mathematics in ECTS	Mathematics in student hours	Notes
Armenia (ASPU)	30	13	GROUP 1 BSc Engineering programs	42	1260	Specialization—Informatics
Armenia (ASPU)	30	13	GROUP 2 BSc Engineering programs	32	960	Specializations—Physics and Natural Sciences
Armenia (ASPU)	30	13	GROUP 3 BSc Technology and Entrepreneurship, Chemistry	9	270	Specialization—Technology and Entrepreneurship, Chemistry
Armenia (ASPU)	30	13	GROUP 4 BSc Psychology and Sociology	9	270	Specializations—Psychology and Sociology
Armenia (NPUA)			GROUP 1 BSc Engineering programs	28		Faculty of Applied Mathematics and Physics
Armenia (NPUA)			GROUP 2 BSc Engineering programs	18		All remaining faculties

Table 10.3 Mandatory mathematics in Georgian engineering BSc programs

Country	One ECTS as student hours	One ECTS as contact hours	BSc program	Mathematics in ECTS	Mathematics in student hours	Notes
Georgia (ATSU)	25	12	GROUP 1 BSc Engineering programs	35	875	Faculty of Exact and Natural Sciences. BSc Informatics
Georgia (ATSU)	25	12	GROUP 2 BSc Engineering programs	32.5	812.5	Faculty of Technical Engineering
Georgia (ATSU)	25	12	GROUP 3 BSc Engineering programs	15	375	Faculty of Technical Engineering
Georgia (ATSU)	25	12	GROUP 4 BSc Engineering programs	10	250	Faculty of Technical Engineering
Georgia (BSU)	25	9	GROUP 1 BSc Engineering	10	250	Programs of Civil Engineering; Transport; Telecommunication; Mining and Geoengineering
Georgia (BSU)	25	9	GROUP 2 Architecture	5	125	Program of Architecture
Georgia (BSU)	25	9	GROUP 3 Computer Science	5	125	Program of Computer Science

Georgia (GTU)	27	12	GROUP 1 BSc Engineering programs	15	405	Faculties of Power Engineering and Telecommunications; Civil Engineering; Transportation and Mechanical Engineering; Informatics and Control Systems; Agricultural Sciences and Biosystems Engineering
Georgia (GTU)	27	12	GROUP 2 BSc Business-engineering programs	15	405	Business-Engineering Faculty
Georgia (GTU)	27	12	GROUP 3 BSc	10	270	Faculties of Architecture; Mining and Geology; Chemical Technology and Metallurgy)
Georgia (GTU)	27	12	GROUP 4 BSc	10	270	International Design School
Georgia (UG)	25–27	6.5	GROUP 1 BSc	24	600–648	Informatics (+elective courses)
Georgia (UG)	25–27	6.5	GROUP 2 BSc	30	750–810	Electronic and Computer Engineering (+ elective courses)
Georgia (UG)	25–27	6.5	GROUP 3 MSc	12	300–324	+ elective courses

Some differences may be detected from the tables. The amount of ECTS varies between engineering BSc programs from 10 ECTS to 75 Russian CUs, which is about 100 ECTS. The contact hours per ECTS are between 6.5 and 13. The time students use in studying engineering mathematics is different between the universities. If all the universities would like to fulfill the SEFI 1 Core, then some universities are resourcing less time for teaching and learning. This is unfortunate, as the quality of learning depends strongly on the amount of time spent on teaching and learning. This is not the only criteria, but one of the important criteria. As mathematics plays an essential role of engineering education it should have a sufficient role in engineering BSc curricula and it should be resourced to be able to reach its goals.

The major observations from the national curricula are the following:

- Russian courses cover more topics and seem to go deeper as well. The amount of exercise hours seems to be larger than EU. The overall number of credits is comparable, but the credits are different (1 cr = 36 h (RU), 1 ECTS = 25–30 h (EU)) therefore per credit, more time is allotted to Russian engineering students for studying mathematics. The amount of mandatory mathematics varies with BSc programs between 7–75 CU (9–100 ECTS). The medium is 20 CU, which is about 27 ECTS. This means that coverage of the SEFI topical areas varies with BSc program. The highest exceeds well the SEFI Core 1, but the lowest lacks some parts. The medium of mandatory mathematics among the programs (20 CU ≈ 27 ECTS) is close to European universities.
- In Armenia, the amount of engineering mathematics in engineering BSc programs varies from 42 ECTS to 18 ECTS. The contents of engineering mathematics is very much the same as in EU but Armenian courses must cover more topics for 18 ECTS than the EU-universities (27–40 ECTS). The amount of lecture/exercise hours/ECTS is about the same as in the EU and the overall number of credits are well comparable.
- In Georgia, the amount of engineering mathematics varies from 35 to 10 ECTS in engineering BSc programs. The minimum 10 ECTS is low compared with the comparable EU, Russian and Armenian degree programs. The coverage of engineering mathematics courses is still very much the same as in EU. This means that there is not as much time for teaching/studying as in other universities. This may reflect negatively to students' outcome competencies. In some cases Georgian credits seem to be higher for the same amount of teaching hours.

10.4 Course Comparison

For course comparison, each partner university selected 1–3 courses, which were compared with similar courses from the EU. In the SEFI classification, the selected courses are the key courses in engineering education. They are taught mostly on BSc and partly on MSc level. In general, the BSc level engineering mathematics courses should cover contents described by SEFI in Core 0 and Core 1. Core 0 contains essentially high-school mathematics, but it is not necessarily on a strong footing or studied at all in schools in all the countries at a level of mastery. The topical areas of Core 1 may vary, depending on the engineering field. Engineering mathematics curriculum may contain elective mathematics courses described in SEFI Core 2 or Core 3. Depending on engineering curricula, these courses can be studied at the BSc or MSc level.

Courses on the following topical areas were selected for comparison between the EU and Russia:

- Engineering Mathematics, Mathematical Analysis
- Discrete Mathematics, Algorithm Mathematics
- Algebra and Geometry
- Probability Theory and Statistics
- Optimization
- Mathematical Modeling

Courses on the following topical areas from the Georgian and Armenian universities were compared with EU universities:

- Engineering Mathematics, Mathematical Analysis
- Calculus
- Discrete Mathematics, Algorithm Mathematics
- Linear Algebra and Geometry
- Probability Theory and Statistics
- Mathematical Modeling

As mathematics is a universal language, the contents of the courses were always in the SEFI core content areas. The course comparison shows that the contents of the courses are comparable. Sometimes a direct comparison between courses was not possible because the topics were divided in the other university between two courses and thus single courses were not directly comparable.

In most of the courses the didactics was traditional and course delivery was carried out in the same spirit. The teacher gives weekly lectures and assignments related with the lectures to the students. The students try to solve the assignments before or in the tutorials or exercise classes.

Students' skills are assessed in exams. The typical assessment procedure may contain midterms exams and a final exam or just a final exam. In the exams students solve examination problems with pen and paper. The teacher reviews the exam papers and gives students their grades. Sometimes student's success in solving assignments or their activity during class hours was taken into account. In some Georgian universities there were tendencies to use multiple choice questions in the exam.

The use of technology to support learning was mainly at a developing stage, and the way it was used depended very much on the teacher. In some universities learning platforms or learning management systems like LMS Blackboard or Moodle was used. Mathematical tool programs (MATLAB, R, Scilab, Geogebra) were known and their use rested much on the teacher's activity.

10.5   Results and Recommendations

10.5.1   Course Development

The contents of the engineering mathematics courses is very much the same in the EU and Russian and Caucasian universities. However, in the EU engineering mathematics is more applied. In other words Russian and Caucasian students spend more time learning theorems and proofs, whereas European students study mathematics more as an engineering tool. We recommend changing slightly the syllabus and instruction from "theorem-to proof" style by putting more emphasis on applications. Topics, applications, examples, related to engineering disciplines, should be added to improve engineering student's motivation to study mathematics.

Traditionally mathematics has been assessed by pen and paper types examinations. Students' assessment could be enhanced so that it covers new ways of learning (project works, essays, peer assessment, epistemic evaluation etc.). Multiply choice questions may be used to give feedback during the courses, but replacing final exams by multiple choice questions cannot be recommended.

10.5.2   Use of Technology

Mathematical tool programs (Sage, Mathematica, Matlab, Scilab, R, Geogebra etc.) are common in EU in teaching and demonstrating how mathematics is put into practice. These programs are known in Russia, Georgia and Armenia, but their use could be enhanced to solve modeling problems from small to large scale. The use of e-Learning (e.g. Moodle for delivery and communication, Math-Bridge as an intelligent platform for e-learning), could be increased in the future to support students' independent work and continuous formative assessment.

10.5.3   Bridging Courses

In the EU, the practices for bridging/remedial courses have been actively developing in the last several decades. With the shift to Unified State Exam and the abolishment of preparatory courses for school abiturients, Russian universities lack the mechanisms to prepare upcoming students to the requirements of university-level math courses. There is also a lack of established practices for bridging courses in the Georgian and Armenian universities to prepare upcoming students to the requirements of university-level math courses. With the shift to Standardized SAT Tests, this becomes problematic, as many students enroll in engineering studies without even a real math test, hence with incorrect expectations and low competencies. Moreover, the needs differ from one student to the next and bridging courses have to be individualized and adapted to specific purposes.

10.5.4   Pretest

Many universities have level tests on Core 0 level for enrolling students to gain an understanding of the mathematical skills new students have and do not have. This makes it possible to detect the weakest students and their needs, providing them further specific support from the beginning of their studies. Math-Bridge system might be a valuable tool here.

10.5.5   Quality Assurance

Quality assurance is an important part of studies and development of education in the EU. Student feedback from courses should be collected and analyzed, as well as acceptance rates, distribution of course grades, and the use of resources in Russian and Caucasian universities.

A necessary, but not sufficient, principle to guarantee the quality of mathematics education is that mathematics is taught by professional mathematicians. This is one of the cornerstones which, in addition to mathematics being an international science, make degrees and courses in mathematics comparable all over the world.

Reference

1. SEFI (2013), A Framework for Mathematics Curricula in Engineering Education. (Eds.) Alpers, B., (Assoc. Eds) Demlova M., Fant C-H., Gustafsson T., Lawson D., Mustoe L., Olsson-Lehtonen B., Robinson C., Velichova D. (http://www.sefi.be).